AF251339

INTRODUCTORY STATISTICS
A Service Course

Other books by A. H. Pollard:

An Introduction to the Mathematics of Finance
Pergamon Press Australia, 1968

Demographic Techniques
(with F. Yusuf and G. N. Pollard)
Pergamon Press Australia, 1974

INTRODUCTORY STATISTICS
A Service Course

A. H. POLLARD

Professor of Economic Statistics
Macquarie University

Pergamon Press Australia

Pergamon Press (Australia) Pty Limited, 19a Boundary Street, Rushcutters Bay, NSW 2011
Pergamon Press Ltd, Headington Hill Hall, Oxford OX3 OBW
Pergamon Press Inc.. Fairview Park, Elmsford, NY 10523

© 1972 A. H. Pollard
First edition published in Australia 1968 by
Pergamon Press (Australia) Pty Limited under the title of
A Service Course in Statistics

Second edition, revised and enlarged 1972
Printed in Hong Kong by Dai Nippon Printing Co. (H.K.) Ltd.
Reprinted 1972, 1975

National Library of Australia Card Number and SBN 08 017352 7
Library of Congress Catalog Card Number 72-189085

Contents

Preface

This book was prepared for a short series of lectures delivered to first year students at Macquarie University. The purpose of these lectures was to introduce statistical ideas and concepts to students whose special interests and abilities lay in other fields but who required some knowledge of statistical reasoning and techniques. Most of the students were majoring in psychology, earth sciences, economics or accounting; however significant numbers of students majoring in english and history also attended, because they were enrolled for one of the former subjects. The lectures were converted to book form chiefly to assist the off-campus students of the university who prepare themselves for the examinations solely by private study.

With this background it is clear that severe restrictions were placed on the contents of the book:

1. Little mathematical knowledge could be assumed; little more than an elementary knowledge of graphs, of simple algebraic manipulation and of logarithms. It was found necessary to include in the *Appendices* a simple introduction to the use of logarithms for arithmetical calculations.

2. The time which students could devote to the subject was severely limited — a problem with all service courses — and the temptation to add material to the course had to be resisted.

3. Topics of special interest to certain groups but of limited interest to others — index numbers and seasonal adjustments in the case of economists, factor analysis in the case of psychologists etc. — are, at Macquarie University, dealt with in specialist courses and are therefore not included here.

4. The presentation had to be at a level to meet the needs of the off-campus student who has no tutorial or other teaching assistance.

It was felt that, with these various restrictions, the aim of presenting as much information as possible, in a simple form and in a short time, would best be achieved by adopting a problem-solution type of approach. The treatment is therefore heavily weighted in favour of examples, selected from many fields. Full solutions are provided for the group of exercises which appear at the end of each chapter. Some miscellaneous problems are also included at the end of the book and for these answers only are given. Specimen examination papers are also included.

For the good student and for the average student we have found that the material contained in the book is adequate to enable them to understand the basic assumptions and to apply the elementary techniques to routine problems. For weaker students and those with little arithmetical competence, however, we have found it necessary to provide long lists of exercises for practice in such routine operations as calculating deviations, using normal distribution tables, the summation notation, logarithms and the like. It was not considered appropriate to include such supplementary material in this book. It is intended however to publish it separately in the form of a workbook for students who need such exercise.

While the book has been prepared specifically for the course entitled 'Introductory Statistics' at Macquarie University, it is hoped that it will also be found useful to the many students with similar backgrounds who experience some difficulty with the subject.

I should like to express my sincere thanks to Mr. E. H. Oliver, for his constructive criticisms of the text and for the valuable advice which he so freely gave. This was particularly valued because of his considerable experience in conducting courses at this level. My thanks are also due to Miss Helen Knight for an excellent typing job from a most difficult manuscript.

A.H.P.
Macquarie University
November, 1970

1

The Presentation of Data

What is statistics?

Over the centuries society has solved many of its problems on a simple trial and error basis. Decisions were often based on hunches. With the development of a scientific approach to problems it has become increasingly common to collect the facts in numerical form and to analyse this numerical information. Figures, rather than hunches, have become the usual basis for decisions.

Figures, however, have to be correctly understood and correctly handled and this is the aim of the subject called 'statistics'. It has grown rapidly in importance during the last couple of decades and is still growing in importance. Progress in the field of statistics has had, and will continue to have, a marked effect on the progress made in both physical and social sciences, and thus, indirectly, affects the rate of human progress.

The word 'statistics' is used in several different senses. It is used to describe the mass of figures themselves, as well as the methods used for describing or presenting the figures and the methods by which conclusions are drawn from the figures.

It has been said that anything can be proved by statistics. D. Huff[1] (1954), wrote a book, *How To Lie With Statistics*. In some practical arguments and in some advertising material these techniques have been used, deliberately I suspect, to good effect. It is most important in a democracy that the man in the street should have some facility in reading and understanding figures, and no doubt for this reason the subject of statistics has at long last found its way into the school syllabus.

[1]   D. Huff, *How to Lie with Statistics*, Gollancz, London, 1954

We are not, however, commencing a study of this subject so that we can deal with advertisers or politicians. Introductory statistics has been set down as a compulsory subject for students studying a number of disciplines because it is not possible to assess the results of experiments in these fields without some understanding of statistical language, statistical techniques, and statistical ideas. In fact, in most disciplines at a university, students will at some stage encounter some form of statistical presentation or analysis. Even in the most unexpected places statistics appear. Let us select for consideration one of the more unlikely fields of statistical enquiry.

The authorship of early Greek prose

In the days when books were written by hand, the rights of authorship were limited and what would now be considered forgery was common.

The names of reputable authors were borrowed for many reasons. Soon after Socrates died, his son complained that the booksellers were putting his father's name on any rubbish, as the name sold the book. Another common motive for substitution was prestige. Views which would receive scant attention under the name of the begetter would, at least, get a critical review if they bore the name of Plato or Aristotle or some leading authority in the field. Another practice, not yet extinct, was for the head of some institution such as the medical school of Hippocrates, to have his name on all work which emerged from the school.

As a result no early Greek authorship can be taken at face value; it must be justified by critical study. It is rare to have conclusive evidence of authorship in contemporary history and hence a subjective analysis of style plays an important part. This, however must depend to some extent on the analyst.

Here the statistician has made a valuable contribution. If we take the works written without doubt by, say, Xenophon and calculate the distribution of sentence length, the frequency of words like *kai* (and), *de* (but), the frequency with which *de* begins a sentence and so on, we have numerical measures of the author's style. Spurious works can usually be rejected by these methods.

The method has, however, to be used with caution; for instance if some pages are purely a literal quotation from someone else these must be excluded.

One of the most celebrated cases of disputed authorship concerns the fourteen Epistles of the New Testament from *Romans* to *Hebrews*. Some credit Paul with writing all of them, some credit him with none; others take a middle view. A. Q. Morton[1] (1965), following this procedure, showed that 'If Paul is defined as the author of *Galatians*, then he also wrote *Romans, Corinthians I and II*. The remaining Epistles come from at least six hands.'

In discussions on Morton's paper at the Royal Statistical Society the (presumably facetious) suggestion was made that Paul may have had one or more secretaries to deal with his correspondence, a most unlikely situation in those days. Sometimes he may have written his own letters, sometimes dictated them and sometimes said, 'You know the answer to that one, Miss Wallinger.'

Tables

You will have noticed from reading technical journals or from reading the daily press that statistical information is frequently presented in the form of tables. It is a psychological fact that figures presented higgledy-piggledy are far harder to understand than figures presented in an orderly fashion. The first step to take, therefore, after figures have been collected is to set them out in an orderly, readily comprehensible fashion.

Regardless of its source any set of data can be presented in several ways, some less effective than others. For instance, consider the following data on birth trends, extracted from publications of the Australian Commonwealth Statistician. We present it in narrative form, that is in ordinary language, and then in tabular form. It will be obvious which form is the more clear and concise.

1. Narrative form

Statistics for Australian confinements in 1965 show that out of 83,584 nuptial confinements to mothers under twenty-five, 42,707 resulted in single male births, 40,189 in single female births and 688 in multiple births. For mothers aged twenty-five and over in the same year the respective figures were 121,554; 61,431; 58,567 and 1,556. The 65,846 nuptial confinements in 1955 to mothers under twenty-five produced

[1] A. Q. Morton, 'The authorship of Greek prose,' *Journal of the Royal Statistical Society*, A128, 1965, p.169

33,632 single male births, 31,647 single female births and 567 multiple births. In this earlier year the respective figures for mothers aged twenty-five and over were 131,071; 66,194; 63,140 and 1,737. In the case of ex-nuptial confinements, there were in 1965, 10,429 in all to mothers under twenty-five, consisting of 5,412 single male births, 4,931 single female births, and 86 multiple births. The corresponding 1965 figures for mothers aged twenty-five and over were 4,945; 2,486; 2,386 and 73 respectively. In 1955 to mothers under twenty-five the total number of ex-nuptial confinements was 4,460 made up of 2,293 single male births, 2,143 single female births and 24 multiple births. For mothers aged twenty-five and over in that year the corresponding figures were 3,988; 1,989; 1,931 and 68.

Table 1.1

The number of nuptial and ex-nuptial confinements in Australia in 1965 and 1955

	Nuptial		Ex-nuptial	
	under 25	25 and over	under 25	25 and over
1965 confinements				
Single Male	42,707	61,431	5,412	2,486
Single Female	40,189	58,567	4,931	2,386
Multiple	688	1,556	86	73
TOTAL	83,584	121,554	10,429	4,945
1955 confinements				
Single Male	33,632	66,194	2,293	1,989
Single Female	31,647	63,140	2,143	1,931
Multiple	567	1,737	24	68
TOTAL	65,846	131,071	4,460	3,988

Source: *Demography Bulletin,* 73 and 83, Bureau of Census and Statistics, Australia

Advantages of tabular layout

1. It enables the required figures to be located quickly.

2. It enables comparisons to be made more easily.

3. It reveals patterns within the figures which cannot be seen in narrative form (e.g. the proportion of confinements which are ex-nuptial has clearly increased over the 10 years).

4. It takes up less space.

You will agree it is still difficult to obtain directly from this table the answer to such questions as: Has the sex ratio changed? Is it different for mothers of different ages? Is it different for ex-nuptial births? Are multiple confinements becoming more frequent? Is the increased incidence of ex-nuptial confinements mainly in the younger age group? and so on. This information can more readily be seen if, for example, appropriate percentages are calculated.

How to build a table

1. Ask yourself the purpose of the table and see how this purpose is most clearly achieved.
2. Decide which are main headings and which sub-headings.
3. Decide whether totals are sufficient or whether percentages or proportions would give more fruitful information.
4. Combine unimportant figures with others or eliminate them.
5. Make the table compact and simple.
6. Give it an exact precise title.
7. State the source of the data.
8. State the units clearly.
9. Check that headings to columns and to rows are not ambiguous.
10. Show totals, sub-totals and percentages, as required.

Graphs

We have shown how tables, properly planned and labelled, and including percentages if required, make data easier to understand. It is a familiar fact, however, that most people understand a visual presentation more clearly than a verbal or a numerical one. In presenting data therefore we can effect a further improvement by representing the data visually — that is, in the form of graphs or diagrams.

For example, the number of persons (in thousands) in Australia receiving unemployment benefit as at the end of each month during 1965 and 1966 is shown in *Table 1.2*.

Table 1.2

Number of persons receiving unemployment
benefit in Australia (,000)

Year	Jan	Feb	March	April	May	June	July	Aug	Sept	Oct	Nov	Dec
1965	15.8	13.7	12.0	12.3	12.6	12.7	12.1	11.0	10.0	9.6	10.5	17.6
1966	19.6	17.6	16.8	17.6	18.1	19.1	19.2	18.6	17.6	16.1	15.9	26.3

Source: *Reserve Bank of Australia, Statistical Bulletin,* Feb. 1966 and Feb. 1967

Clearly the numbers were greater in 1966 than in 1965. But can you tell, at a glance from the table, whether there is any trend in these figures?

Now look at the same information shown in graphical form in *Fig. 1.1:*

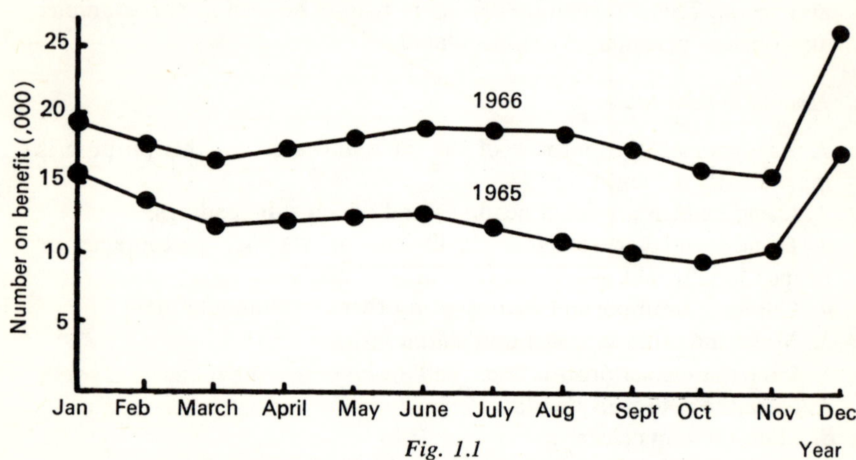

Fig. 1.1

Persons receiving unemployment benefit in Australia

Source: *Reserve Bank of Australia, Statistical Bulletin,* Feb. 1966 and Feb. 1967

The seasonal trend is obvious and the relative movement of one year against the other can clearly be seen.

A graph shows visually the value which one quantity (the dependent variable) assumes for a given value of the other quantity (the independent variable).

Remember

1. The graph must have a clear and concise title.

2. The independent variable should be placed on the horizontal axis.

3. The vertical scale should always start at zero.

4. Axes should be clearly labelled.

5. Curves must be distinct.

6. A graph must not be overcrowded with curves.

7. The source of the data must be given.

But most important of all

8. The correct impression must be given. Because graphs are based on visual representation they are open to every trick in the field of optical illusion. Look at the two graphs in *Figs. 1.2* and *1.3* which have been taken from Moroney, *Facts From Figures*[2]. They represent the same data but give very different impressions. Modest progress can be made to look phenomenal by stretching the vertical scale.

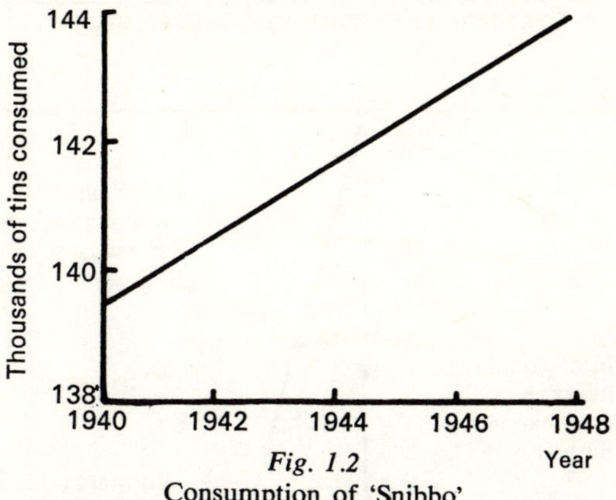

Fig. 1.2
Consumption of 'Snibbo'

When the vertical scale is 'stretched' by commencing the scale at some value well above zero, the rise in consumption seems phenomenal.

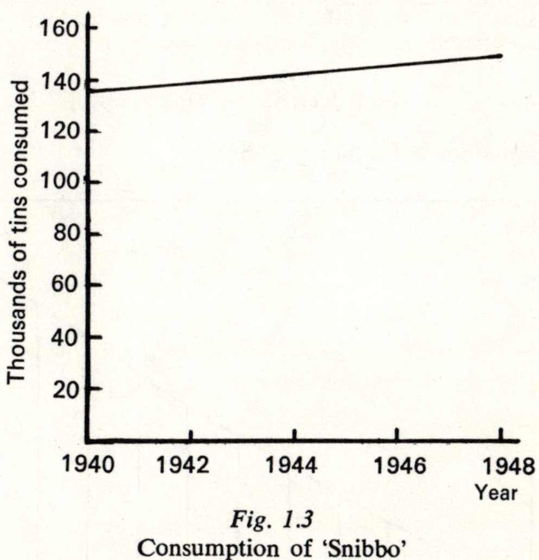

Fig. 1.3
Consumption of 'Snibbo'

[2] M. J. Moroney, *Facts from Figures,* Penguin, England, 1951, p.28. Reprinted with kind permission of the author.

Fig. 1.3 shows the modest progress actually made. Now look at *Fig. 1.4* which shows the advertiser's graph from *Facts From Figures*.   Note that it is devoid of scales.

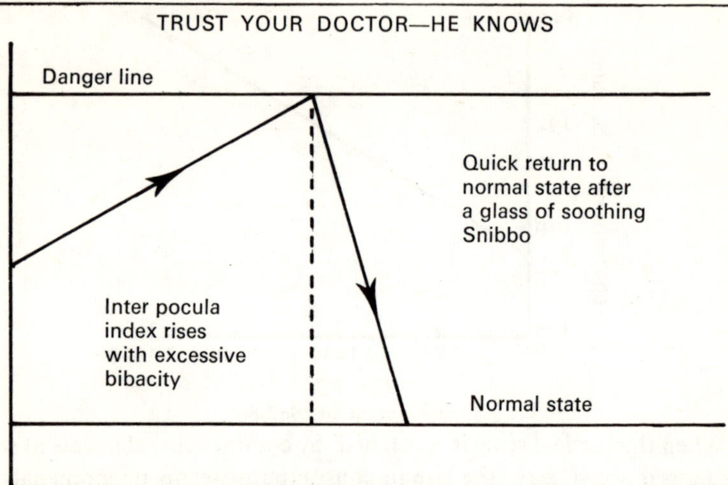

After a consumption of alcohol, your 'Inter Pocula Index' rises to what may prove a dangerous level, with serious risk of muscular atony. In such cases the taking of a therapeutic nostrum has untold effect as a sedative and restorative. There is no finer nostrum than Snibbo.

Fig. 1.4
The Advertiser's Graph

Fig. 1.5 presents some further graphs:

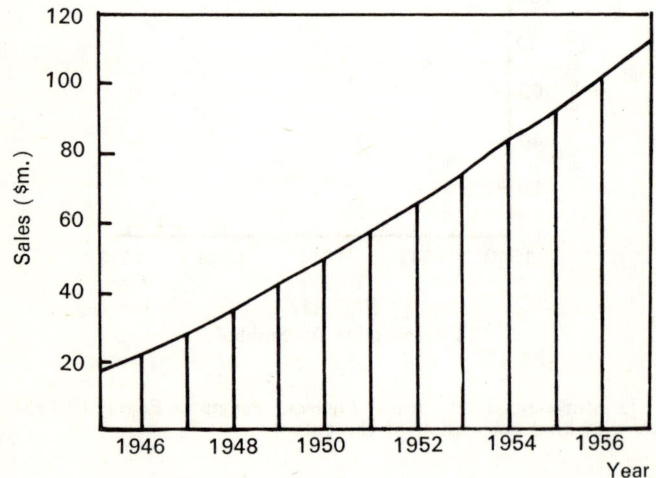

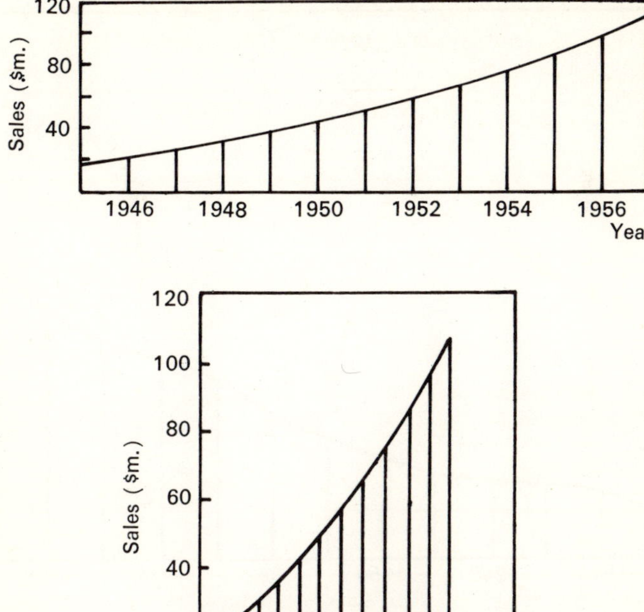

Fig. 1.5
Life assurance sales by the XYZ Company

The three graphs represent the same data with different scales, all vertical scales starting from zero. However they give very different impressions. *Fig. 1.6* is probably a fair choice of scale for this data.

A false impression can also be created by varying the vertical scale. *Figs. 1.7* and *1.8* represent the same data but give different impressions. *Fig. 1.7* is the usual form of presentation and gives the correct impression of growth; *Fig. 1.8* uses a logarithmic scale which is not appropriate here but is regarded as standard in some contexts.

A time series (of which *Figs. 1.1* to *1.8* are examples in the form of graphs) is the name given to data which shows how one or more items vary over a period of time.

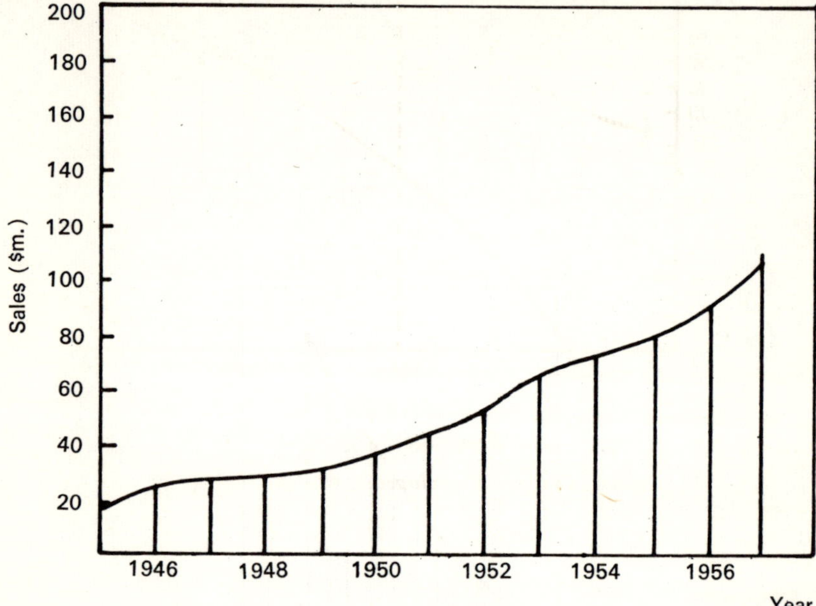

Fig. 1.6
Life assurance sales by the XYZ Company

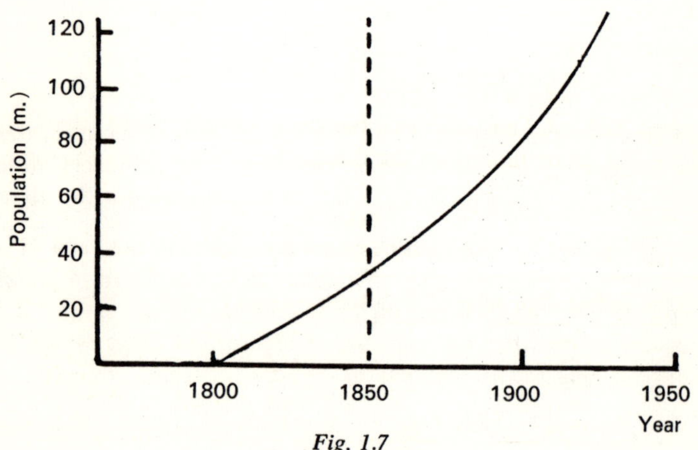

Fig. 1.7
Population of USA
Source: *Collier's Encyclopedia,* vol. 19, 1961, p. 11

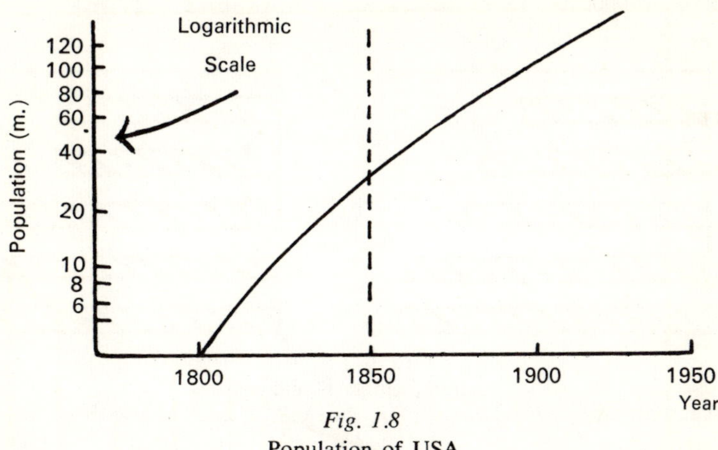

Fig. 1.8
Population of USA
Source: *Collier's Encyclopedia*, vol. 19, 1961, p. 11

Diagrams

Graphs are not the only way of presenting data visually. The other
forms — of which there are many — we shall call diagrams. Here are
some examples.

Simple bar charts

The value of Australia's imports and exports in millions of dollars is
shown in *Table 1.3* and also as a bar chart in *Fig. 1.9*. Most people find
that the bar chart illustrates the fluctuations better than the table.

Table 1.3
Australia's imports and exports

Year ending 30th June	Imports ($m.)	Exports ($m.)
1962	1770	2155
1963	2163	2152
1964	2373	2783
1965	2905	2651
1966	2940	2721

Source: *Year Book of Australia*, 1968, Bureau of Census and Statistics, no. 54,
p. 341

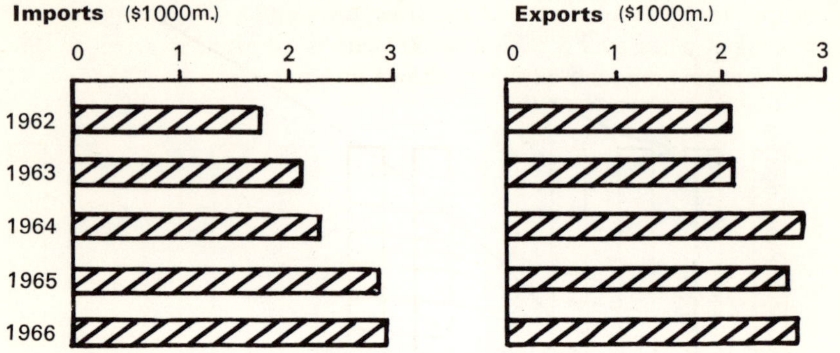

Fig. 1.9
Australia's imports and exports
Source: *Year Book of Australia,* 1968, Bureau of Census and Statistics, no. 54,
p. 341

Compound bar charts

The population of Australia in total, and split into three age groups of
economic significance, is shown in *Fig. 1.10* in the form of a compound
bar chart.

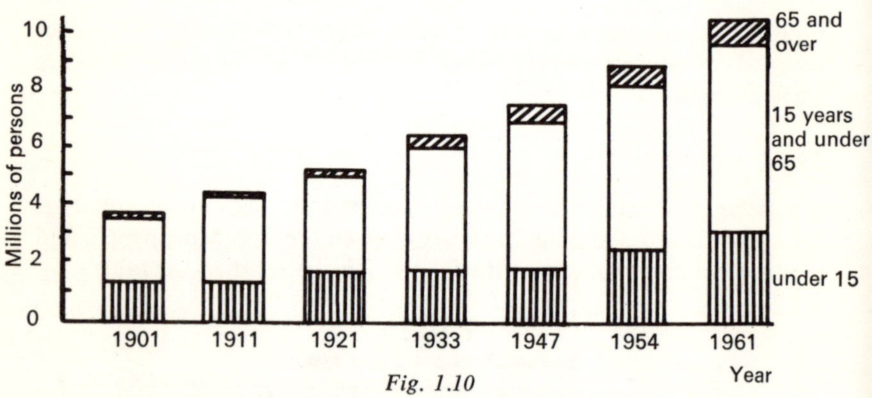

Fig. 1.10
The population of Australia in three age groups

Source: *Year Book of Australia,* 1968, Bureau of Census and Statistics, no. 54,
p. 134

Percentage compound bar charts

These consist of bars all of the same height but divided into parts
representing the proportions having certain characteristics; e.g. the

proportion of families of different sizes in Australia. The change which has taken place over a period of 60 years is clearly illustrated by the percentage compound bar chart shown in *Fig. 1.11*.

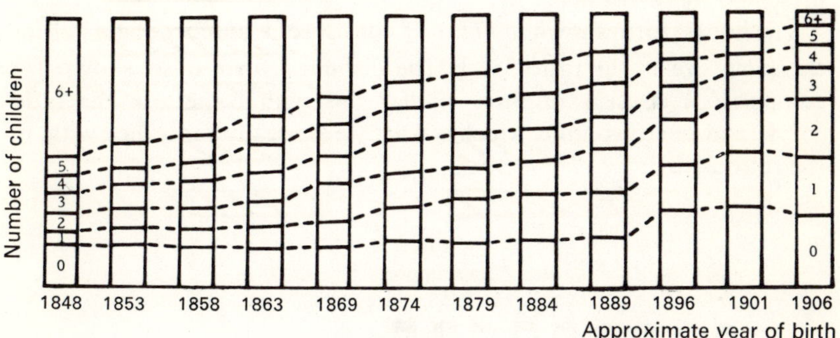

Fig. 1.11
Distribution of family size (wives over 45 at census)

Source: A. H. Pollard and G. N. Pollard, *'Fertility in Australia'*, *Transactions of the Institute of Actuaries of Australia and New Zealand*, 1966, p.19.

Multiple bar charts

These consist of a series of two or more bar charts. An example is given in *Fig. 1.12*.

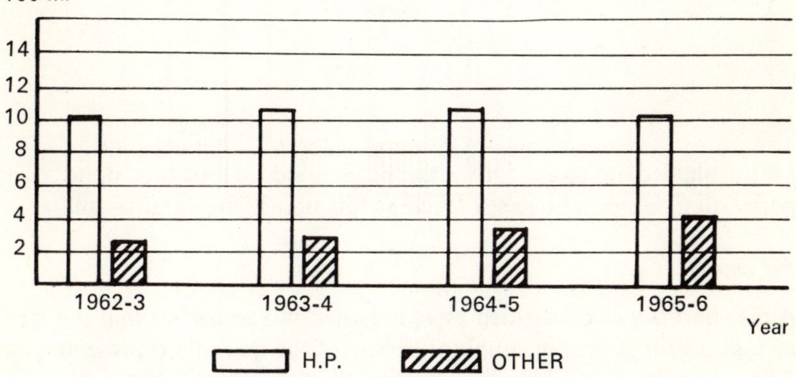

Fig. 1.12
Instalment credit for retail sales — balances outstanding
Source: *Year Book of Australia*, 1968, Bureau of Census and Statistics, no. 54, p. 706

Pictograms or pictographs

These resemble a bar chart but are made up of pictures in order to attract attention, for instance, sales of three brands of petrol, represented by petrol tins.

The honest form shown in *Fig. 1.13* consists of a number of tins all of the same size in the ratio 3:2:1. The dishonest form (also shown for comparison) consists of three similar tins with heights in the ratio 3:2:1, and hence volume, which is what we are really concerned with, in the ratio 27:8:1.

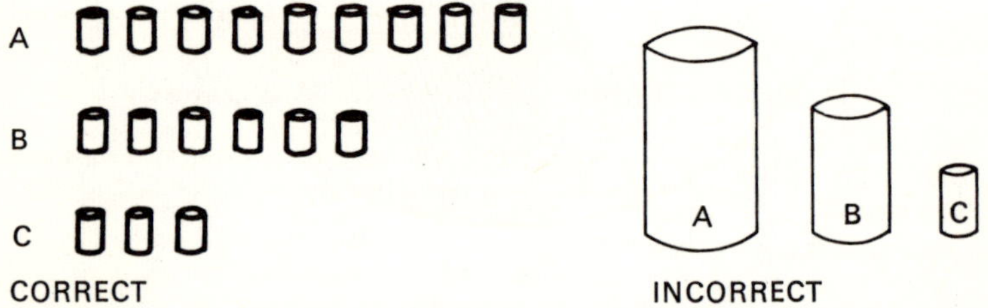

Fig. 1.13
Sales of petrol by companies A, B and C

Maps

Maps, shaded in some form, are often used to indicate density of population, height above sea level, production figures and so on.

Pie charts

A pie chart is a circle divided by radial lines into sectors so that the area of each sector is proportional to the size of the quantity represented by that sector. It is useful for representing the proportions that make up a given total. Two or more pie charts of different sizes should not be used for representing two or more sets of figures as pie charts of different sizes are difficult to compare.

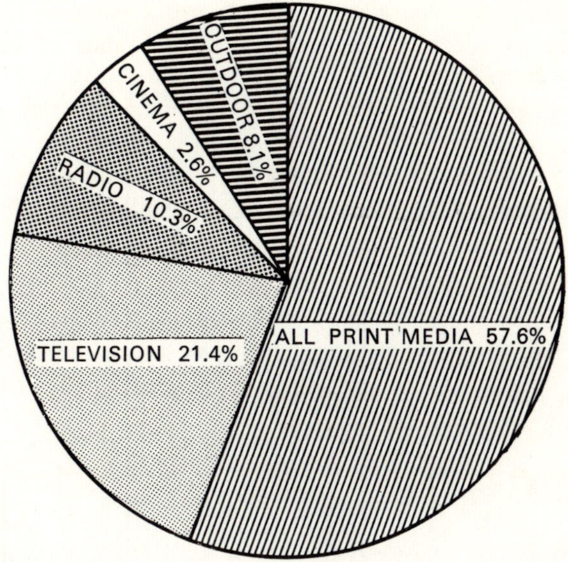

Fig. 1.14

1964 Australian advertising expenditure divided between the main media

Source: *Australian Economy,* W. D. Scott & Co., Sydney, 1966

General comments

When writing a paper it is well worth while spending a great deal of time in determining the most convincing visual form in which your statistics can be presented. There is much scope for ingenuity.

Frequently both positive and negative figures occur, e.g. imports and exports, emigration and immigration, percentage change over previous year, profit and loss. In such cases bar charts on opposite sides of a line, perhaps coloured differently, may be used, as in *Fig. 1.15*.

Colour can also be used to good effect in multiple bar charts, pie diagrams etc. Density of population can be shown on maps by varying intensities of, say, grey or red.

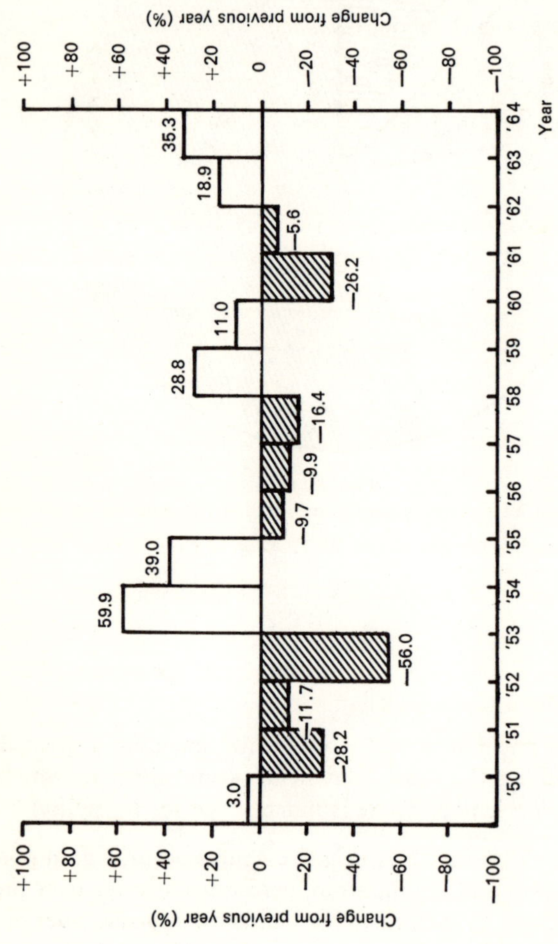

Fig. 1.15

Net permanent and long term migration to Australia

Source: *Australian Economy*, W. D. Scott & Co., Sydney, 1966

Exercises

The following Exercises *1.1* to *1.6* are not on the work of this chapter but are to test whether the student is familiar with some of the simple mathematical processes necessary for later chapters. A knowledge of the use of logarithms is required for calculation work and a simple treatment of the subject is therefore included in *Appendix II.* (page 224).

1.1 Given $A = 2$, $B = -3$, $C = 4$ and $D = -2$, calculate

a. $3A - 4B$

b. $6A + 5B + 2 - 4D$

c. $\dfrac{AB + CD}{AC - BD}$

d. $A^2 + B^2 - 4CD$

e. $3(A + 2C) - 2(B - 4D)$

f. $\dfrac{B^2 - C^2}{BC - 1}$

g. $\sqrt{2A^2 - B^2 + C^2}$

h. $\sqrt{\dfrac{2A^2}{B^2} + \dfrac{3C^2}{D^2}}$

1.2 Draw a graph of $Y = X^2 - 2X - 5$ for values of X from -3 to $+5$.

1.3 The present values of a 10 year annuity at various rates of interest are shown in the following table:

Rate (% per annum)	3	4	5	6	7	8
Present value	8.530	8.111	7.722	7.360	7.024	6.710

Draw a graph of these results and estimate from it the value of a 10 year annuity at $6\frac{1}{4}\%$ per annum.

1.4 If $L = -2$, $M = 3$, $N = \frac{1}{2}$, $P = -\frac{1}{3}$ and $R = \frac{1}{6}$, calculate

a. $\sqrt{\dfrac{(M - P)^2}{R} + \dfrac{(L - M)^2}{N}}$

b. $\dfrac{R - P}{\sqrt{L^2 + M^2}} (L^2R[N + P])$

1.5 Use logarithms to calculate

a. $\dfrac{(0.05632)\ (579.2)}{(12.32)\ (0.003875)}$

b. $\sqrt{\dfrac{(46.84)\ (0.00752)^3}{(2.264)^5}}$

c. $\dfrac{39.42}{0.2364}\sqrt{\dfrac{(52.61)\ (0.07461)}{(0.02865)\ (0.6829)}}$

1.6 If

$$
\begin{aligned}
X_1 &= 2 & Y_1 &= 5\\
X_2 &= 4 & Y_2 &= -3\\
X_3 &= 3 & Y_3 &= 1\\
X_4 &= -2 & Y_4 &= -3\\
X_5 &= 6 & Y_5 &= 4
\end{aligned}
$$

find

a. $\displaystyle\sum_{1}^{5} X_r$ f. $\displaystyle\left(\sum_{1}^{5} Y_r\right)^2$

b. $\displaystyle\sum_{1}^{5} Y_r$ g. $\displaystyle\sum_{1}^{5} X_r Y_r$

c. $\displaystyle\sum_{1}^{5} X_r^2$ h. $\displaystyle\sum_{1}^{5} X_r^2 - \tfrac{1}{5}\left(\sum_{1}^{5} X_r\right)^2$

d. $\displaystyle\left(\sum_{1}^{5} X_r\right)^2$ i. $\displaystyle\sum_{1}^{5} Y_r^2 - \tfrac{1}{5}\left(\sum_{1}^{5} Y_r\right)^2$

e. $\displaystyle\sum_{1}^{5} Y_r^2$ j. $\displaystyle\sum_{1}^{5} X_r Y_r - \tfrac{1}{5}\left(\sum_{1}^{5} X_r\right)\left(\sum_{1}^{5} Y_r\right)$

Note: Students not familiar with the Σ notation should read the brief explanation of it on page 21.

The following two problems are on the work of Chapter 1:

1.7 Draw
a. a graph
b. a bar diagram
c. a pictogram of the population of Australia from the figures in the following table:

Year	1905	1915	1925	1935	1945	1955	1965
Population (m.)	4.03	4.97	6.00	6.76	7.43	9.31	11.48

1.8 The following table sets out the heights in centimetres of 100 mature plants of a certain type:

Height (cm)	Number of plants
61 — under 62	1
62 — under 63	4
63 — under 64	5
64 — under 65	7
65 — under 66	13
66 — under 67	15
67 — under 68	24
68 — under 69	10
69 — under 70	7
70 — under 71	7
71 — under 72	4
72 — under 73	1
73 — under 74	2

Draw a bar diagram of this information.

Determine the number of plants which have heights less than 62, 63, 64 . . . etc. centimetres and draw a graph of the results. Assuming these 100 plants are typical of all mature plants of this type determine from the graph the height which would be exceeded by

a. One half of all mature plants of this type

b. One quarter of all mature plants of this type

c. Three quarters of all mature plants of this type.

What proportion of mature plants of this type would be between 66 cm and 70 cm in height?

2

Frequency Distributions

Raw data

This is the term used for the original figures which, for example, are recorded as an experiment or survey progresses.

Table 2.1, for example, shows the number of calls made by 80 insurance men in the course of one week.

Table 2.1

Number of calls made by 80 insurance men in one week

184	204	168	110	188	142
172	149	115	153	112	132
171	140	154	161	157	175
125	138	114	105	195	138
173	200	152	123	184	142
144	176	109	150	146	187
207	217	202	164	162	178
174	156	153	140	159	148
155	155	210	177	120	167
152	149	179	138	166	155
117	213	132	159	192	149
135	118	121	162	130	136
122	145	158	134	127	199
161	151				

Source: Fictitious

When presented in this form it is difficult to glean from this mass of information anything more than the fact that the number of calls is approximately 100 to 200 per week. Statistical techniques however may be used to extract the full information contained in these figures. This means that a routine procedure can be followed so that the significant features may be brought out.

Array

If we rearrange the figures in order of size from the smallest to the largest, then such a list of figures is called an array.

Table 2.2 shows the same data in the form of an array.

Table 2.2

Number of calls made by 80 insurance men in one week

105	130	145	155	166	184
109	132	146	155	167	187
110	132	148	155	168	188
112	134	149	156	171	192
114	135	149	157	172	195
115	136	149	158	173	199
117	138	150	159	174	200
118	138	151	159	175	202
120	138	152	161	176	204
121	140	152	161	177	207
122	140	153	162	178	210
123	142	153	162	179	213
125	142	154	164	184	217
127	144				

We can immediately tell from the array that the number of calls ranges from 105 to 217 and that 150 odd is roughly 'in the middle'. Over 200 calls is clearly uncommon, only six insurance men being in this group.

Frequency distribution

The number of occurrences of any event is described as its frequency. You will note that 153 calls is recorded twice, that is its frequency is two. We could list the different numbers of calls and opposite each number record its frequency, i.e. the number of times it occurs. Such a tabulation is called a frequency distribution. The sum of the frequencies must obviously equal the total number of items in the raw data.

The sigma notation is often used to represent the sum of a number of successive items. Thus the sum of the frequencies of occurrence of from 140 calls to 150 calls inclusive is written $\sum\limits_{140}^{150} f$ which from *Table 2.3* equals twelve. Σf without limits stands for the total frequency.

Table 2.3 shows the same data in the form of a frequency distribution.

Table 2.3
Number of calls made by 80 insurance men in one week

Number of calls	Frequency	Number of calls	Frequency	Number of calls	Frequency	Number of calls	Frequency
105	1	134	1	155	3	177	1
109	1	135	1	156	1	178	1
110	1	136	1	157	1	179	1
112	1	138	3	158	1	184	2
114	1	140	2	159	2	187	1
115	1	142	2	161	2	188	1
117	1	144	1	162	2	192	1
118	1	145	1	164	1	195	1
120	1	146	1	166	1	199	1
121	1	148	1	167	1	200	1
122	1	149	3	168	1	202	1
123	1	150	1	171	1	204	1
125	1	151	1	172	1	207	1
127	1	152	2	173	1	210	1
130	1	153	2	174	1	213	1
132	2	154	1	175	1	217	1
				176	1		

Total frequency $= 80$

Grouping

There are still too many figures to enable us to grasp the information effectively. The data may be simplified by grouping the figures into classes. For example, there are eight insurance men who recorded from 100 to 119 calls. That is, in this class there is a frequency of eight. If we divide the number of calls into groups and record the frequency of occurrences in each group we obtain a grouped frequency distribution. The result in this case is shown in *Table 2.4*.

Table 2.4
Number of calls made by 80 insurance men in one week

Number of calls (x)	Frequency (f)
100 — under 120	8
120 — under 140	15
140 — under 160	26
160 — under 180	17
180 — under 200	7
200 — under 220	7
	80

N.B. Σf must equal 80

From this grouped frequency distribution we can detect a pattern in the figures. The number of calls clusters around the 140 — under 160 class. This simplification brings out the general pattern, but, of course, a certain amount of information is lost. We do not know where, within the range 100 — under 120, the eight cases lie. Hence we have simplification at the expense of a certain amount of exactness. We would have lost less information had we chosen groups 100 — under 110, etc.

Class limits must be clearly stated; 100 — 120 is ambiguous; 100 to under 120, however, makes it clear into which group 120 is placed.

The number of classes should be carefully chosen. With too many classes (e.g. twenty) it is difficult to grasp the information. With too few classes (e.g. three) the result is inexact.

Class intervals should be equal wherever possible and should be in round figures like 5 or 10 (not 7 or 11). Classes should be chosen where possible so that occurrences within classes balance roughly around the mid-point of the classes.

In practice it is usual to proceed directly from the raw data to the grouped frequency distribution. This can be carried out simply as follows:

1. Select the class intervals.

2. Take each figure from the raw data and draw a unity (1) opposite the appropriate class in the table and at the same time tick the figure in the raw data. After four unities have been entered in a class, record the fifth by a line drawn through the previous four and then leave a blank, as this simplifies totalling.

3. Check that Σf equals the overall number of observations in the raw data. This overall number is often denoted by n.

Table 2.5 illustrates the procedure in the case of the 80 insurance men.

Table 2.5

Compiling a grouped frequency distribution

Class interval		Frequency
100 — under 120	卌 ⏐⏐⏐	8
120 — under 140	卌 卌 卌	15
140 — under 160	卌 卌 卌 卌 卌 ⏐	26
160 — under 180	卌 卌 卌 ⏐⏐	17
180 — under 200	卌 ⏐⏐	7
200 — under 220	卌 ⏐⏐	7
		$\Sigma f = \overline{\underline{80}}$

Histogram

A frequency distribution can be presented visually. A common form of representation is the histogram which is like a simple bar chart in which the area of a bar represents the frequency of occurrences within the class interval represented by the width of that bar. *Fig. 2.1* shows the histogram for the frequency distribution in the case of the 80 insurance men.

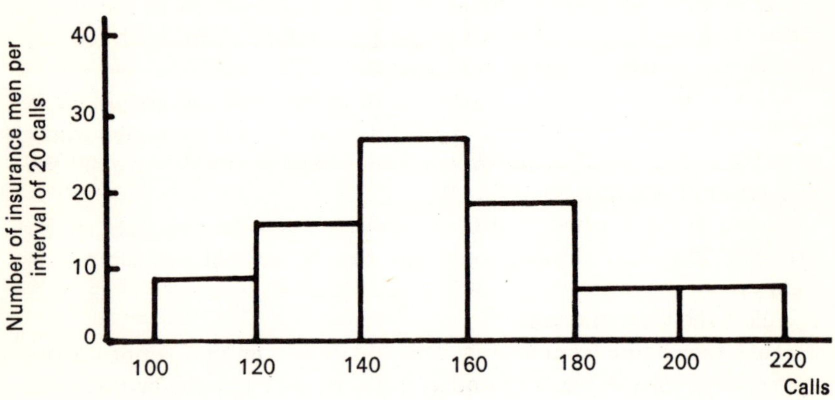

Fig. 2.1

Histogram showing the distribution of insurance men by number of calls made

It is, we repeat, the *areas* of bars which represent the frequency; so that if the class intervals are unequal the height of a particular bar should be reduced in the proportion that its class interval is increased. The height, in our example, is the number *per class interval of 20 calls* and the *area* is the frequency for the class interval used. If a larger class interval is chosen, the frequency has to be divided by the increase in size of this larger class interval, to give the height. This is illustrated, in *Fig. 2.2* where 100 — under 140 is used as a class interval.

The number of men who made less than 160 calls per week is represented by the areas of the three left bars of the histogram in *Fig. 2.2*. The number is apparently a little more than half of the 80 men.

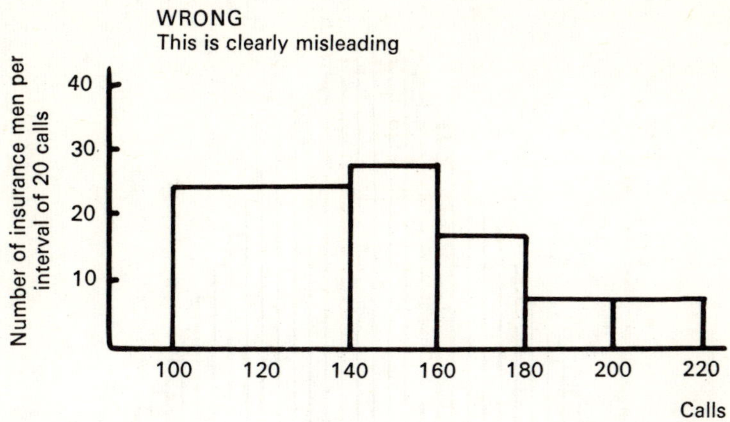

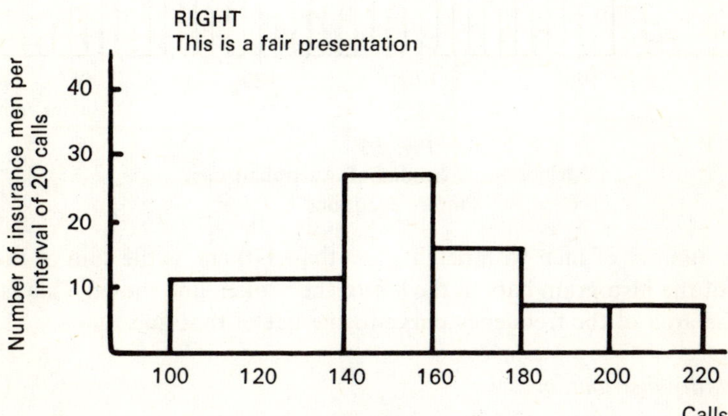

Fig. 2.2

Distribution of insurance men by number of calls made using unequal class intervals

Frequency curve

If we had such a large amount of data that very small class intervals could be used, then the top of the histogram would get very close to a smooth curve. If, for example, the heights of all men in Australia could be recorded in centimetres, then with such very fine intervals the histogram would approach a smooth curve. This smooth curve is called a frequency curve.

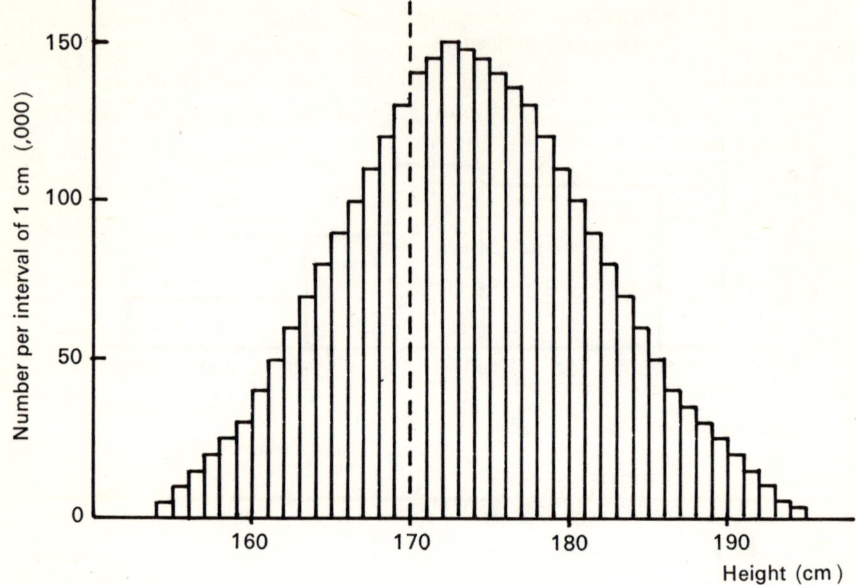

Fig. 2.3
Distribution of heights of Australian men
Source: Fictitious

The number of men with heights less than 170 cm. is the sum of the areas of the histogram bars to the left of the broken line shown. That is, it is the area of the frequency curve to the left of that line.

Probability distribution

We are usually more interested to know what *fraction* of the men has heights less than (say) 170 cm rather than the actual number. If therefore we make the total area under the frequency curve equal to unity, then the area to the left of (say) 170 cm gives the fraction of men who are less than 170 cm in height. Such a curve is called a probability distribution curve and will often be encountered in future work.

Frequency polygon

If, instead of using a bar for each class interval as in the case of histograms, we put a dot at the same height over the mid point of the class interval and join these dots we obtain a frequency polygon. It is usual to include also the first class interval at each end which has no occurrences in it, (i.e. frequency = 0) and indicate this zero frequency by a dot at the centre of the class interval on the horizontal axis. This completes the frequency polygon. *Fig. 2.4* shows a frequency polygon for the 80 insurance men.

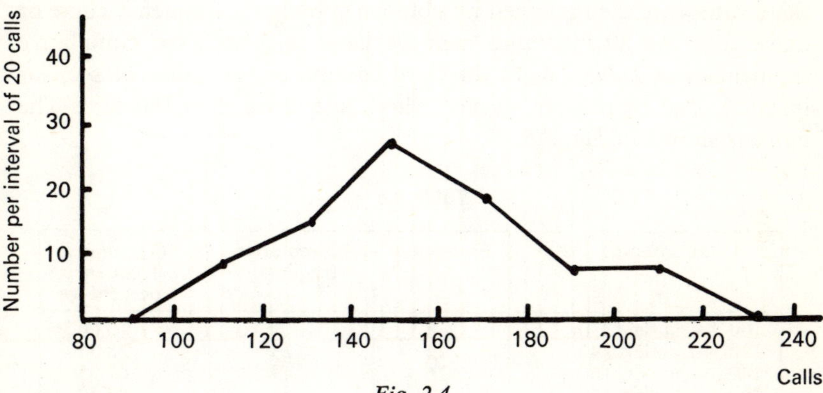

Fig. 2.4

Frequency polygon showing the distribution of insurance men by number of calls

If we draw the frequency polygon and the histogram on the same graph, as is done in *Fig. 2.5*, it can be seen that their areas are equal.

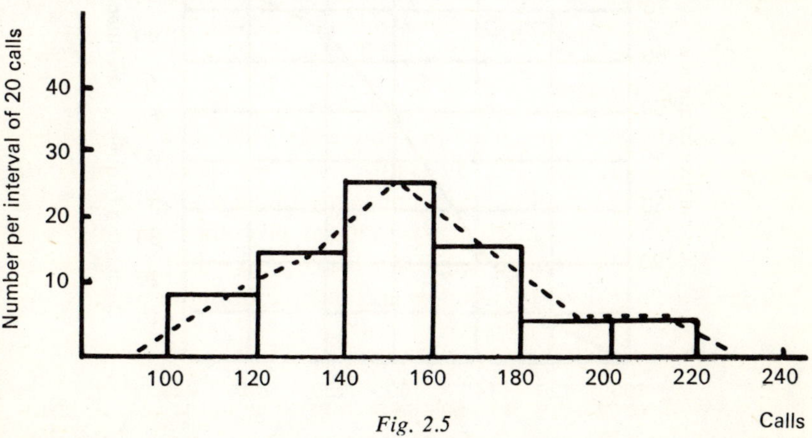

Fig. 2.5

Histogram of *Fig. 2.1* and frequency polygon of *Fig. 2.4* drawn on the same graph

Ogive

We are often interested, not in the number of cases in a class interval, but the number of cases *below* a given value. We may like to know the number of men who made less than 160 calls or less than 180 calls etc. To obtain this readily we need to add the frequencies cumulatively. If these values are then graphed we obtain a cumulative frequency curve or ogive. For the 80 insurance men we have calculated the cumulative frequencies in *Table 2.6*. In the third column of this table we see, for instance, that there were 49 men who made fewer than 160 calls. The ogive is shown in *Fig. 2.6*.

Table 2.6

Class interval	Frequency	Cumulative frequency	Cumulative frequency percentage
100 — under 120	8	8	10.00
120 — under 140	15	23	28.75
140 — under 160	26	49	61.25
160 — under 180	17	66	82.50
180 — under 200	7	73	91.25
200 — under 220	7	80	100.00
	80		

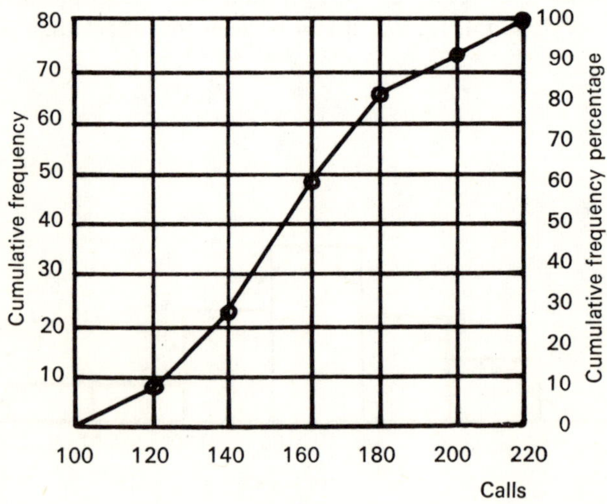

Fig. 2.6
Ogive for the distribution of insurance men by number of calls made

The ogive in *Fig. 2.6* is called a '*less than*' ogive because it is based on the number of men who made *less than* a certain number of calls. We could have calculated the number of men who made *more than* a certain number of calls and plotted the data. The result would have been a '*more than*' ogive.

We are usually more interested in the *proportion* of cases less than a given value, e.g. the proportion of the insurance men who made less than 180 calls per week. These values have also been calculated in *Table 2.6*. It is interesting to note that the same ogive can represent this information also if the vertical scale is adjusted so that the total frequency represents 100. This new scale is shown on the right hand side of *Fig. 2.6*.

If we wish to compare the number of calls made by men from companies whose number of employees differ, then this comparison can easily be made by calculating the cumulative frequency percentage.

There are thus two ways of determining the fraction or percentage of observations which are less than a given value. Referring to the illustration in *Fig. 2.3*, the percentage of men with heights less than 170 cm can be determined

either (*1*) by finding the area *to the left of 170 cm* on the probability distribution curve;

or   (*2*) by finding the height *at 170 cm* on the corresponding ogive.

Exercises

2.1 The heights in centimetres jumped by 60 primary school students were

67	60	53	67	63	71
63	56	49	63	64	70
49	53	60	57	61	65
57	53	50	60	50	51
67	68	58	54	61	66
58	54	50	65	68	48
46	72	66	69	65	58
70	46	62	51	47	62
48	59	58	62	55	52
52	56	63	59	56	56

Find *a.* The greatest height
    b. The lowest height
    c. The range of heights
    d. The height jumped by the 8th highest student
    e. The heights of the top four jumps
    f. How many students jumped above 63 cm.
    g. How many students jumped 53 cm. or less
    h. What percentage of students jumped above 54 cm but not above 64 cm.

2.2 Assuming that the heights in the previous question were measured to the nearest centimetre, draw up a frequency table using as class boundaries $45\frac{1}{2}$ cm , $49\frac{1}{2}$ cm , $53\frac{1}{2}$ cm , etc.
a. Determine from this table
   (i) The class interval
   (ii) The lower limit of the third class
   (iii) The frequency of the third class
   (iv) The relative frequency of the third class expressed as a percentage
   (v) The class interval with the largest frequency
   (vi) The percentage of students who jumped above $57\frac{1}{2}$ cm.
   (vii) The percentage who jumped more than $57\frac{1}{2}$ cm but less than $65\frac{1}{2}$ cm.

b.   *(i)* Draw the histogram
   (ii) Draw up a relative frequency distribution table

(iii) Draw the relative frequency polygon
(iv) Draw up a cumulative frequency distribution table
 (v) Draw up a percentage cumulative frequency distribution table
(vi) Draw a less than ogive
(vii) Draw a more than ogive.

c. From these last two ogives estimate the number of jumps that were
   (i) less than 67 cm.
  (ii) greater than 53 cm.
 (iii) between 53 cm and 67 cm.

2.3 The numbers of male deaths in Australia in 1967 in age groups were
as follows:

Ages	Number of deaths
30 — under 35	164
35 — under 40	231
40 — under 45	390
45 — under 50	628
50 — under 55	1008
55 — under 60	1712

Draw a less than ogive and estimate from it the number of deaths in
Australia in 1967 of males aged from 37 to under 52.

3

Averages

What are they? Why are they used?

So far we have endeavoured to extract the important information contained in a mass of figures by visual presentation. This is, however, not always possible or desirable. The manager of the company employing these 80 insurance men should be watching and keeping a record of the calls made by his staff each week. He would need to have the weekly record extending over a couple of years in order to detect unfavourable trends. If the record were kept in histogram form it would be like a bulky family photograph album. What is required is one single figure which represents for this purpose all the figures for the week. Such a single representative figure is called an *average*.

Arithmetic mean

It might be thought that the total number of calls might be the appropriate figure. If, however, the company is expanding and the staff increasing the total figure does not help. A suitable figure would be the total calls by all the men divided by the number of men. This is called the arithmetic mean, or simply the mean. From the raw data the mean number of calls by the 80 men during the week being studied was 155.4.

Sportsmen will be familiar with the method adopted for recording a cricketer's performance during a season. 'Total runs' is an unfair comparison because batsmen may not have had the same number of opportunities. A listing of individual scores against each batsman's name would produce a mass of confusing figures. Hence the arithmetic mean is used. In this context it is called the 'batting average'.

The arithmetic mean is written $\bar{x}$ and the formula to obtain it is

$$\bar{x} = \frac{\Sigma x}{n}$$

where Σx is the sum of the observations (in our case the sum of the number of calls) and n is the number of observations (in our case the number of insurance men).

If a particular value of x occurs several times, as is the case when data is presented in the form of a frequency table, to obtain Σx it is shorter to multiply each value of x by the number of times it occurs i.e. by its frequency f, rather than adding all the values of x. In this case the formula for $\bar{x}$ may be written

$$\bar{x} = \frac{\Sigma fx}{\Sigma f}$$

because n the number of observations equals Σf.

We shall use these formulae to calculate the mean number of calls of the 80 insurance men. From *Table 2.1* on page 20,

$$n = 80$$
$$\Sigma x = 184 + 172 + 171 + \ldots + 136 + 199 = 12432$$

Hence $\bar{x} = \dfrac{12432}{80} = 155.4$ calls.

Applying the second formula to the same data presented in the form of a frequency table, namely *Table 2.3* on page 22,

$$\Sigma f = 1 + 1 + 1 + \ldots + 3 + 2 + 2 + \ldots + 1 + 1 = 80$$
$$\Sigma fx = 1 \times 105 + 1 \times 109 + \ldots + 3 \times 138 + \ldots$$
$$\ldots + 1 \times 217 = 12432$$

Hence $\bar{x} = \dfrac{12432}{80} = 155.4$ calls, as before.

Using an arbitrary origin

It is possible to simplify the numerical work by a simple device. All the men made 100 calls; some made a few more, some many more. If we record the number of calls made in excess of 100 the raw data becomes

$$84, 72, 71, 25, 73 \ldots \ldots \ldots \ldots 49, 36, 99.$$

The mean of these figures is 55.4 calls (in excess of 100). Hence the mean of the original data

$$= 100 + 55.4 = 155.4 \text{ calls, as before.}$$

We could have obtained smaller figures still by working from 150 calls as a base. Then the number of calls (above or below 150) would be

$$34, 22, 21, -25, 23, -6 \ldots$$

The mean of these figures is 5.4 calls (in excess of 150). Hence the mean of the original data

$$= 150 + 5.4 = 155.4 \text{ calls, as before.}$$

The nearer we choose our 'base' to the arithmetic mean the smaller the figures involved in the calculation become.

Hence the first step in calculating the arithmetic mean of raw data is to make a rough guess at it (to some round figure like 150). The second step is to work out the deviations from this figure, which is called an arbitrary origin. The third step is to calculate the mean of these deviations. The mean of the raw data is obtained by adding this latter figure to the arbitrary origin. Thus, if we write d for deviations from the arbitrary origin, then

$$\bar{x} = \text{arbitrary origin} + \frac{\Sigma d}{n}$$

This device is often useful with manual calculations. With computers or calculating machines, however, this simplification is unnecessary.

Arithmetic mean of a grouped frequency distribution

Much labour is saved if we determine the arithmetic mean from a grouped frequency distribution instead of from the raw data. In some cases, information is only available in this form. It is usual to assume then that all the occurrences in a class interval take the value of the mid point of that class interval. The contribution of that class interval to the total (i.e. to Σx) equals the number of occurrences in that class interval multiplied by the value of the mid point of that class interval.

Hence the calculation is as follows:

Class interval	Mid-value (m)	Frequency (f)	f times m
100 — under 120	110	8	880
120 — under 140	130	15	1,950
140 — under 160	150	26	3,900
160 — under 180	170	17	2,890
180 — under 200	190	7	1,330
200 — under 220	210	·7	1,470
		$\Sigma f = 80$	12,420

$$\therefore \bar{x} = \frac{12,420}{80} = 155.25 \text{ calls}$$

This differs slightly from the true mean computed from the raw data (155.4 calls). The difference, which is negligible for practical purposes, arises from the assumption that all the values in a given class assume the value at the mid point of the class. This error is due to the loss of information on grouping.

The calculation can be further simplified by working from an arbitrary origin (say 150).

Class interval	Deviation of mid-value from 150 (d)	Frequency (f)	f times d
100 — under 120	-40	8	-320
120 — under 140	-20	15	-300
140 — under 160	0	26	0
160 — under 180	20	17	340
180 — under 200	40	7	280
200 — under 220	60	7	420
		$\Sigma f = 80$	420

$$\therefore \bar{x} = 150 + \frac{420}{80} = 155.25 \text{ calls}$$

One further simplification can be made; that is to change the unit of deviation to some simple value for calculation purposes only and allow for the changed units in the final line. In our example the units could have been dozens of calls, 10 calls etc. Had we been dealing with lengths, the units could have been feet, yards, miles etc. Some choices would involve much more numerical work than others. Thus if in this problem we make a unit for d equal to 20 calls, we have

Class interval	d	f	f times d
100 — under 120	—2	8	—16
120 — under 140	—1	15	—15
140 — under 160	0	26	0
160 — under 180	1	17	17
180 — under 200	2	7	14
200 — under 220	3	7	21
		$\Sigma f = 80$	$+21$

$$\therefore \ \bar{x} = 150 + 20 \times \frac{21}{80} = 155.25 \text{ calls}$$

Note: The calculation has thus been simplified by

1. Using a grouped frequency distribution

2. Selecting an arbitrary origin

3. Selecting a suitable unit for deviations.

Averaging means

If a cricketer had a batting average of 50 last season and 80 this season we cannot find his average over the two seasons without more information. We do not need to know the individual scores; all we need to know is the number of innings he had each season. Thus if there were 6 innings last season and 5 this season his total last season must have been 50 times 6 or 300, and this season 80 times 5 or 400. This makes a total of 700 over the two seasons from 11 innings, or an average of $63\frac{7}{11}$ runs per innings. Summarizing, his batting average over the two years is

$$\frac{\text{Total score}}{\text{Total innings}} = \frac{50 \times 6 + 80 \times 5}{6 + 5} = \frac{700}{11} = 63\frac{7}{11} \text{ runs}$$

Note: It is not $\dfrac{50 + 80}{2}$ as is often supposed.

Similarly if in a particular 'common' examination paper the 10 students in one class have an 'average' (i.e. arithmetic mean) of 60 marks and the 40 students in a second class have an 'average' of 50 marks, and the 20 students in a third class have an 'average' of 40 marks, then the 'average' for the year (i.e. for the three classes) is

$$\frac{10 \times 60 + 40 \times 50 + 20 \times 40}{10 + 40 + 20} = \frac{3400}{70} = 48\frac{4}{7} \text{ marks.}$$

This ability to average means is a most useful property.

Median

If the data is arranged in order of magnitude, that is in the form of an array, then the median is the value of the middle item in the array.

If there is an odd number of items n, then the median is the value of the $\dfrac{n + 1}{2}$th item. If n is even there is no middle value. For example, with the 80 insurance men the median value would be somewhere between the number of calls of the 40th and 41st man. It is usual when n is even to take for the median the mean of these two values (which are usually nearly equal to one another). That is, it is usual to consider the median as the value of the $\dfrac{n + 1}{2}$th item whether n is even or odd.

Thus the median for the insurance men data

$$= \frac{153 + 154}{2} = 153.5 \text{ calls.}$$

The median is that value in an array which divides it so that there is an equal number of items on either side of it. In the case of a histogram, or of a frequency distribution, or of a probability distribution, the median divides the area under the curve into two equal parts.

Calculating the median from grouped data

In the case of grouped data we can tell in which group the median item lies but we cannot tell its value. It is usual to estimate the value by assuming that the occurrences are evenly spaced within the group. Let us consider the grouped distribution of the insurance men. We require the value of the $40\frac{1}{2}$th item. It obviously lies in the 140 — under 160 group since there are 23 items below this group and 26 in it.

The 23rd item lies below 140 calls and the 24th item above it. If we assume the items to be evenly spaced we may say that 140 calls corresponds to an imaginary $23\frac{1}{2}$th item. Similarly 160 calls corresponds to an imaginary $49\frac{1}{2}$th item.

As the length of the class interval was 20 with 26 items in that interval, the interval between successive items within the group could be estimated as one twenty-sixth of 20, i.e. $\dfrac{20}{26}$. Thus as the $23\frac{1}{2}$th item was 140, the $24\frac{1}{2}$th would be $140 + \dfrac{20}{26}$, the $25\frac{1}{2}$th would be $140 + 2 \times \dfrac{20}{26}$, and so on. The $40\frac{1}{2}$th would thus be

$$140 + 17 \times \frac{20}{26} = 140 + 13.1 = 153.1 \text{ calls.}$$

This result could be obtained more directly by condensing the process into one of simple proportion, as

$$140 + \frac{40\frac{1}{2} - 23\frac{1}{2}}{49\frac{1}{2} - 23\frac{1}{2}} \times (160 - 140) = 153.1 \text{ as before.}$$

The median, from the grouped data, is thus 153.1 calls. The difference between this value and that obtained from the raw data is due to the loss of detail from grouping.

Quartiles

The median is useful if, for example, we wish to separate primary school children into two secondary schools in equal numbers according to their IQ s. The median is the IQ value which makes the separation into equal halves.

The principle can be extended. If we wish to divide them in equal numbers to send to four schools, then the three IQ marks which effect this separation are called the 1st, 2nd and 3rd quartiles. The second quartile is obviously the median and hence the former name is not used.

Let us compute the 1st and 3rd quartiles for the insurance men data. As with the median, it is convenient to identify the quartiles as imaginary observations — the lower quartile being the $\dfrac{n+1}{4}$th observation and the upper quartile the $\left(3 \times \dfrac{n+1}{4}\right)$th observation, the count of observations of course being made on the data arranged in order of magnitude. In this case the quartiles are the values of the $20\frac{1}{4}$th and $60\frac{3}{4}$th items. From *Table 2.2* on page 21 it can be seen that the 1st quartile lies between 136 and 138, say $136\frac{1}{2}$. The 3rd quartile lies between 173 and 174, say $173\frac{3}{4}$.

From the grouped frequency distribution, namely *Table 2.4* on page 22, they may be calculated in the same manner as the median.

The first quartile obviously lies in the 120 — under 140 group. We assume 120 calls lies midway between the 8th and 9th items and hence corresponds to an imaginary $8\frac{1}{2}$th item, and 140 calls midway between the 23rd and 24th items and hence corresponds to an imaginary $23\frac{1}{2}$th item. Thus the number of calls corresponding to the imaginary $20\frac{1}{4}$th item (assuming uniform spacing) is

$$120 + \frac{20\frac{1}{4} - 8\frac{1}{2}}{23\frac{1}{2} - 8\frac{1}{2}} \times 20$$

$$= 120 + \frac{11\frac{3}{4}}{15} \times 20 = 135.7 \text{ calls.}$$

Similarly the third quartile is

$$160 + \frac{60\frac{3}{4} - 49\frac{1}{2}}{66\frac{1}{2} - 49\frac{1}{2}} \times 20 = 173.2 \text{ calls.}$$

The difference between these values and those obtained from the raw data is again due to loss of detail from grouping.

Deciles and percentiles

This principle can be extended to dividing an array into tenths or hundredths, the dividing values being called deciles or percentiles. If we wish to select 18% of pupils on the basis of IQ for a selective school, then the 82nd percentile would be the IQ value which would effect the separation, for we require the *top* 18% of pupils.

It can readily be seen that the simplest method of determining percentiles, deciles and even quartiles is to draw the ogive and then read off the values of the required item from the curve.

Mode

One obvious point of note in a set of data is the value which occurs most frequently. Because this value, the 'mode', need not in any ordinary sense, be typical of the data as a whole, little use is made of it. Data may have several modes; for instance, the insurance men data had three modal values, namely 138, 149 and 155 which each arose three times.

We need not examine methods of estimating the mode from grouped data; it is usually sufficient to identify the modal class; that is the class with the greatest observed frequency.

Mean v. median

The mean treats each item as of equal value and is therefore influenced (or distorted) by extreme values. It is useful in numerical processing and plays an important part in a branch of statistics called statistical inference.

The median is not influenced by extreme values. If one isolated insurance man had made 300 calls, this would be reflected in an increase in the mean but the median would not be affected.

The median is not of the same value as the mean in the more advanced theory but its related quartiles etc. are useful for dividing arrays into equal sections.

Exercises

3.1

   a. Find the mean and median of the following numbers:
$$1, 3, 5, 9, 13$$
   b. Hence write down the mean of the numbers
$$10, 30, 50, 90, 130$$
   c. The following numbers are obtained by adding 51 to the numbers in *a.* Hence write down their mean.
$$52, 54, 56, 60, 64.$$

3.2 A set of 110 Test scores was condensed into a grouped frequency table in two different ways, thus:

Class	Frequency	Class	Frequency
159.5 — under 169.5	1	149.5 — under 169.5	3
149.5 — under 159.5	2		
139.5 — under 149.5	6	129.5 — under 149.5	16
129.5 — under 139.5	10		
119.5 — under 129.5	16	109.5 — under 129.5	36
109.5 — under 119.5	20		
99.5 — under 109.5	19	89.5 — under 109.5	37
89.5 — under 99.5	18		
79.5 — under 89.5	12	69.5 — under 89.5	15
69.5 — under 79.5	3		
59.5 — under 69.5	2	49.5 — under 69.5	3
49.5 — under 59.5	1		
	110		110

Determine
a. The midpoint of the modal class in each case
b. The median in each case
c. The arithmetic mean in each case by the following two methods
   (*i*) By using the original origin
  (*ii*) By selecting an arbitrary origin.

3.3 Find the arithmetic mean of the heights in *Exercise 2.1* by direct addition. Find the arithmetic mean again using 55 as an arbitrary origin. What is the median height?
What is the modal height?

3.4 The profit results of five companies for 1966 are as follows:

Company	Trading Result
A	$175,264 profit
B	$ 4,685 profit
C	$ 14,617 loss
D	$ 7,245 loss
E	$ 1,654 loss

What is the average trading result of the five companies?
Is this average typical of the five companies?

4
Variation

So far we have been trying to simplify a mass of figures by obtaining a single figure — an average — which may be considered to represent the whole. The average is however not the only piece of information which is contained in a set of figures. We are frequently interested in the extent to which individual results vary above and below the average. A knowledge of this variation is often more important than a knowledge of the average value.

A cricketer may have a good average, but is he consistent? Is he very 'up and down'?

If I travel regularly by air to Melbourne for meetings I am not simply interested in the average time of arrival. How often would I be late? In other words, how does the time of arrival vary?

If we are selling packages alleged to contain 2 kg of sugar it may not be good enough to know that the average contained by our packages is 2 kg. If they vary from 1.95 to 2.05 kg this may be in order. If, however, they vary from 1.75 to 2.25 kg we could be in serious trouble.

One student's marks in a number of examinations may average 60, the marks in individual examinations varying from 20 to 100. Another student may also obtain an average mark of 60 but with individual marks varying from 55 to 65. A different interpretation might be placed on these two performances.

In summarizing the results of an experiment or other data we would therefore normally like to indicate

     (1) a representative value or average
and (2) the extent of the variation.

Measures of variation

1. Range

This is the simplest measure and is simply the difference between the largest and smallest values in the sample. For the data in *Table 2.1* (page 20) we obtain

$$\text{Range} = 217 - 105 = 112 \text{ calls.}$$

The range is greatly influenced by extreme values. If one man had made 350 calls the range would become 245 calls, although all other values are unaltered.

2. Quartile deviation

We can get over this difficulty by using the interquartile range. This is the difference between the first and third quartiles. The usual measure however is the quartile deviation (or semi-interquartile range), which is half the interquartile range. Again for the data in *Table 2.1*, using the results on page 39 we obtain

$$\text{Quartile deviation} = \frac{173\tfrac{3}{4} - 136\tfrac{1}{2}}{2} = 18.63 \text{ calls.}$$

3. Mean deviation

Individual values differ from the mean by what is called the 'deviation from the mean' or just the deviation. Consider, for example, the marks obtained by students in a test, set out in *Table 4.1*. The marks total 100 and, as there are 20 students, the mean mark is five.

The deviations of the individual values from 5 (the mean) are given in the second column. They must total zero. If we ignore the signs, i.e. ignore whether the values are greater or less than the mean, and total the deviations irrespective of sign, we get 34 which when divided by the number of marks, namely 20, gives us a value of 1.7 for the *average* deviation irrespective of sign. This is called the mean deviation.

This tells us how far *on the average* individual values are from the mean.

This leads us to the most important measure of variation, namely, the standard deviation.

Table 4.1

Calculation of mean deviation and standard deviation

Mark x	Deviation from mean $x - \bar{x}$	Absolute deviation $\lvert x - \bar{x} \rvert$	Deviation squared $(x - \bar{x})^2$
3	−2	2	4
6	+1	1	1
8	+3	3	9
5	0	0	0
4	−1	1	1
9	+4	4	16
7	+2	2	4
4	−1	1	1
1	−4	4	16
8	+3	3	9
5	0	0	0
3	−2	2	4
6	+1	1	1
2	−3	3	9
5	0	0	0
7	+2	2	4
2	−3	3	9
6	+1	1	1
4	−1	1	1
5	0	0	0
$\Sigma x = 100$	0	$\Sigma\lvert x - \bar{x}\rvert = 34$	$\Sigma(x - \bar{x})^2 = 90$

$$\text{Mean deviation} = \frac{34}{20} = 1.7 \qquad s = \sqrt{\frac{90}{20}} = 2.12$$

4. Standard deviation

If we return to the second column in *Table 4.1* and square each value all the signs become positive. The sum of the squares of the deviations is 90 and the average of the squares of the deviations is therefore 90 divided by 20 or 4.5. If we take the square root of this quantity we obtain the standard deviation (written s) which in this case is 2.12. The formula for the standard deviation, straight from the definition, is therefore

$$s = \sqrt{\frac{\Sigma(x - \bar{x})^2}{n}}$$

The standard deviation is thus the square root of the mean of the squares of deviations from the mean.

An alternative formula for s, which may be shown to be equal to the formula just given, is

$$s = \sqrt{\frac{\Sigma x^2 - n\bar{x}^2}{n}} \quad \text{or} \quad \sqrt{\frac{\Sigma x^2}{n} - \bar{x}^2}$$

As an illustration of the use of this formula, we refer to the data in *Table 4.1*.

$$n = 20$$
$$\Sigma x = 3 + 6 + 8 + 5 + \ldots + 5 \qquad = 100$$
$$\Sigma x^2 = 3^2 + 6^2 + 8^2 + 5^2 + \ldots + 5^2 = 590$$

Hence $\quad \bar{x} = \dfrac{100}{20} = 5 \text{ and } s = \sqrt{\dfrac{590}{20} - 5^2} = \sqrt{4.5} = 2.12$

At first sight it might appear that this is a most complicated method of measuring variation, but it has considerable advantages in later statistical theory.

Standard deviation from a frequency distribution

The data given in *Table 4.1* may be rewritten in the form of a frequency distribution. In this case the square of the deviation must be multiplied by the frequency with which the value occurs. The calculation of s in these cases is shown in *Table 4.2*.

Table 4.2

Calculation of s from a frequency distribution

x	$x - \bar{x}$	f	$(x - \bar{x})^2$	$f \times (x - \bar{x})^2$
1	-4	1	16	16
2	-3	2	9	18
3	-2	2	4	8
4	-1	3	1	3
5	0	4	0	0
6	$+1$	3	1	3
7	$+2$	2	4	8
8	$+3$	2	9	18
9	$+4$	1	16	16
		$\Sigma f = 20$		$\Sigma f(x - \bar{x})^2 = 90$

$$s = \sqrt{\frac{90}{20}} = 2.12 \text{ (as before)}.$$

We chose the above values of f so that the mean came out to be a round figure and hence the deviations were integers. If, as is usual, the mean does not work out as an integer (e.g. for the insurance men data $x = 155.4$ calls) then the deviations are complicated and the above method of calculation is most cumbersome. The standard deviation may be calculated quite simply, however, by using an alternative but algebraically equivalent formula. The formula used to calculate s in *Table 4.2.* is

$$s = \sqrt{\frac{\Sigma f \times (x - \bar{x})^2}{\Sigma f}}$$

An alternative but equivalent formula is

$$s = \sqrt{\frac{\Sigma f x^2}{\Sigma f} - \left(\frac{\Sigma f x}{\Sigma f}\right)^2}$$

Without writing the calculation out in tabular form we obtain, using this formula

$$\Sigma f = 1 + 2 + 2 + 3 + \ldots + 1 \qquad\qquad = 20$$
$$\Sigma f x = 1 \times 1 + 2 \times 2 + 2 \times 3 + 3 \times 4 + \ldots + 1 \times 9 = 100$$
$$\Sigma f x^2 = 1 \times 1^2 + 2 \times 2^2 + 2 \times 3^2 + 3 \times 4^2 \ldots + 1 \times 9^2 = 590$$

Hence $\quad s = \sqrt{\dfrac{590}{20} - \left(\dfrac{100}{20}\right)^2} = \sqrt{4.5} = 2.12$

This is the usual formula for calculating s from a frequency distribution.

The standard deviation is a measure of variation. As this variation clearly remains the same when we add a constant amount to every value of x, we may calculate s using the last formula above, measuring x from some arbitrary origin. If we write d for the values measured from this arbitrary origin the formula for s becomes

$$s = \sqrt{\frac{\Sigma f d^2}{\Sigma f} - \left(\frac{\Sigma f d}{\Sigma f}\right)^2}$$

The purpose of using an arbitrary origin is to reduce the size of the figures in the calculation. This is no great advantage if calculating machines are available.

The use of this formula on the same data is illustrated in *Table 4.3* where $x = 3$ has been chosen as the arbitrary origin.

When dealing with grouped data there is one further simplification which can be introduced. As was the case in calculating the arithmetic mean, we can change the unit of deviation to some simple value and at the end convert our answer back to true units.

We have used this method in *Table 4.4* where s is determined for the insurance men grouped data using 20 calls as a unit for deviation.

Table 4.3

Calculation of s from a frequency distribution using an arbitrary origin

x	d (from 3)	f	$f \times d$	$f \times d^2$
1	-2	1	-2	4
2	-1	2	-2	2
3	0	2	0	0
4	$+1$	3	$+3$	3
5	$+2$	4	$+8$	16
6	$+3$	3	$+9$	27
7	$+4$	2	$+8$	32
8	$+5$	2	$+10$	50
9	$+6$	1	$+6$	36

$$\Sigma f = 20 \qquad \Sigma fd = +40 \qquad \Sigma fd^2 = 170$$

$$\therefore \frac{\Sigma fd}{\Sigma f} = 2 \qquad \frac{\Sigma fd^2}{\Sigma f} = 8.5$$

Hence $s = \sqrt{8.5 - 4} = 2.12$ (as before).

Table 4.4

Calculation of s for the grouped data in *Table 2.3* (page 22)

x	d	f	$f \times d$	$f \times d^2$
100 — under 120	-2	8	-16	32
120 — under 140	-1	15	-15	15
140 — under 160	0	26	0	0
160 — under 180	$+1$	17	$+17$	17
180 — under 200	$+2$	7	$+14$	28
200 — under 220	$+3$	7	$+21$	63

$$\Sigma f = 80 \qquad \Sigma fd = +21 \qquad \Sigma fd^2 = 155$$

$$\frac{\Sigma fd}{\Sigma f} = \frac{21}{80} = 0.2625 \qquad \frac{\Sigma fd^2}{\Sigma f} = \frac{155}{80} = 1.9375$$

$$\sqrt{\frac{\Sigma fd^2}{\Sigma f} - \left(\frac{\Sigma fd}{\Sigma f}\right)^2} = \sqrt{1.9375 - (0.2625)^2} = 1.367$$

Hence $s = 1.367 \times 20 = 27$ calls approx.

Variance

The square of the standard deviation is useful in later work. It is given the name variance (s^2).

Comparison of measures

The range and quartile deviation are simple to calculate and understand but are seldom used. The quartile deviation takes no account of extreme values whereas the range is unduly affected by them. Neither measure takes note of the distribution of individual values.

The mean deviation is easy to understand and gives equal weight to all values. The standard deviation is difficult to understand and, although it weighs extreme values highly (by squaring deviations), it is generally used because it is of great value in later statistical theory.

Units

Range, quartile deviation, mean deviation, standard deviation and all averages (mean, median, mode) are all measured in the same units as the original values.

Skewness

We have now moved one stage further in our endeavour to describe the characteristics of a set of figures by simple measures. We have used

(*1*) Averages — to represent the typical size of the figures;

and (*2*) Measures of dispersion — to indicate the degree or extent of variation.

Two sets of figures can have the same values for (*1*) and (*2*), for example the same mean and standard deviation, and yet be characteristically different.

Look at the three histograms or the three frequency curves in *Fig. 4.1*:

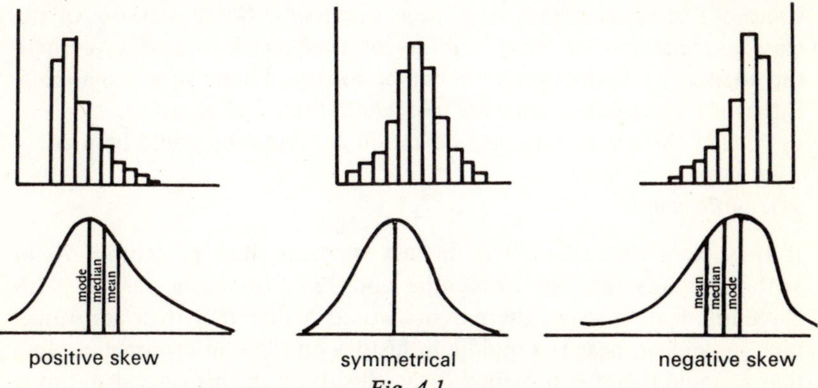

Fig. 4.1
Symmetrical and skewed distributions

They have the same mean and standard deviation and yet there is a characteristic difference. We are referring to the degree of symmetry or balance of the set of figures.

If the peak is in the centre and the curve falls away evenly on either side it is said to be symmetrical. If the peak is to the left and the curve falls away more rapidly to the left than to the right, it is described as having a positive skew or being positively skewed. If the reverse, it has a negative skew.

It can be seen that for a symmetrical distribution the mean, mode and median are all equal and for a skewed distribution the median usually lies between the mean and the mode.

Scores

If a schoolboy came home and told his parents that in a test he scored 39 in history, 76 in arithmetic and 135 in vocabulary and english grammar, his parents could learn nothing from this about his school performance. If they were interested they would immediately ask for each subject in turn such questions as:
What was the top mark?
What was the bottom mark?
Were there many above you?
What was the average mark?   and so on.

They are concerned with his relative performance and they are trying to find out where their son's mark lay in the general pattern of class marks. If they and the teacher had both had statistical training this information could have been conveyed to the parents quite simply using some of the measures we have been discussing. Some measure of the class average and of the variability of class marks would give them the necessary information but it can be conveyed more simply than this.
Either (1) the median, quartile, percentile family of measures
or   (2) the mean, standard deviation combination could be used.

Percentile scores

If they were told that 39 in history was the 48th percentile, 76 in arithmetic was the 29th percentile and 135 in english was the 97th percentile for the class, the parents would realize that their son ranks high in english, near the middle in history and low in arithmetic. Note that a single figure is provided for each subject in this case also, but it conveys a real meaning. A mark of 39 in history may be 39 marks out of 100 or 39%; it may stand for 39 out of 40. It may be the top mark in the class in either case or it may be the bottom mark; we do not know. The 48th percentile however tells us that 48% of the class had marks below this figure. The percentile score does not tell us anything about his actual marks or the percentage of marks gained but it does tell us his relative position in class.

Standard scores

If the parents had been told that the mean score for the class in arithmetic was 82 they would have known their son was below average but no real idea how far below average. This is really only being told that his deviation from the mean mark was —6 (i.e. 76 —82). This is sometimes spoken of as 'deviation score'. Such a performance is obviously not good, but we still need some idea of the variability of the marks to assess just how poor he was relative to the other students. A single quantity which combines the deviation score with this other requirement is the 'standard score' which expresses the individual deviation away from the mean in units of the standard deviation.

$$\text{Thus the standard score} = \frac{\text{deviation from the mean}}{\text{standard deviation}}$$

If the standard deviation of the arithmetic mark was 10 then the son's standard score in arithmetic was —0.6. This single standard score, we shall see later, gives a fairly accurate idea of the son's performance relative to his colleagues.

Exercises

4.1 A cricketer records the following scores in a season:
$$1, 16, 82, 14, 29, 17, 142, 18, 7, 59$$
What is his batting average?
Another cricketer in the same season scored:
$$36, 38, 41, 39, 33, 44, 40, 37, 32, 45$$
What is his average?
Who is the more consistent batsman?

4.2 Here are two examples of vague classification.
Make the classification precise.
a. The diameters in centimetres of trees are distributed as follows:

Diameter	3	4	5	6	7
Frequency	8	12	17	15	3

b. The wages of employees are distributed as follows:

Wages ($)	10 — 12	12 — 14	14 — 16	16 — 18
Frequency	5	12	14	8

4.3 *a.* Find the standard deviation of the following figures:
$$1, 3, 7, 13, 21$$
b. What is their mean deviation?

c. Write down the standard deviation of the following figures:
$$10, 30, 70, 130, 210$$
d. If we were to add 7 to each of the figures in (*c*) what would the standard deviation of the new figures be?

4.4 In a class of 48 the examination marks ranged from 45 to 70 as shown in the following table:

Mark	Number of students with this mark
45	2
50	3
55	20
60	16
65	6
70	1
	—
Total	48

Convert these marks to standard scores (i.e. number of standard deviations from the mean).

4.5 The systolic blood pressures (measured in millimetres of mercury) of 20 healthy individuals under special study were:

132, 115, 122, 142, 124, 136, 118, 128, 128, 130
126, 125, 134, 128, 120, 131, 138, 110, 124, 140

a. Arrange this data in the form of an array
b. Determine the mean, the median, the lower and upper quartiles
c. Calculate the range, the semi-interquartile range, the standard deviation, the mean deviation
d. Using intervals of 5 millimetres draw up a frequency table
e. Draw the histogram and the frequency polygon
f. Compute the relative frequencies
g. Draw an ogive from which you can determine the percentage of individuals who have a systolic blood pressure of
 (*i*) Less than 135 mm
 (*ii*) More than 125 mm
(*iii*) Between 120 and 135 mm.
h. Determine the mean, the median, the upper and lower quartiles again, this time from the frequency table
i. Calculate the standard deviation from the frequency table
j. What percentage of the data lies within one standard deviation of the mean?

4.6 Construct histograms for the data in the following table and superimpose frequency polygons on them:

Marks scored in IQ tests by pupils
at two schools

	Number of pupils	
IQ Mark	School A	School B
75 — under 85	15	43
85 — under 95	25	99
95 — under 105	40	54
105 — under 115	108	40
115 — under 125	92	14
125 and over	20	0
	300	250

a. Calculate the mean and standard deviations of the two distributions
b. Calculate the median IQ and the interquartile range in each case
c. If the top one-fifth from both schools combined were to be selected for transfer to a special school what IQ mark should be adopted for the purpose?
d. If 450 pupils in school C had a mean IQ of 106 what would be the mean IQ of the pupils in schools A, B and C combined?

5

Samples and Populations

If we wished to study the intelligence level of university students in Australia we might commence by obtaining the IQ score of all students at Macquarie University. It would be a risky assumption to assume that the results obtained were fairly representative of Australia. They could be higher, or lower. If however we were to obtain IQ figures for every student in every university in Australia, then from this complete set of figures we could make statements about Australian university student IQs with complete confidence. We could quote the mean and range of variation with exactness. It is frequently not possible or it is too costly to obtain the complete set of IQ figures. We often have to be satisfied with a part, sometimes a small part, of them.

The entire set of figures is called the population and that part which is available to us is called a sample.

The IQ scores of all Macquarie University students would constitute a population with respect to questions about IQs of Macquarie University students at that time. It would, however, be a sample with respect to questions about the IQs of Australian university students.

Populations may be finite or infinite. A finite population is one which has both a first member and a last member. The rats in a certain laboratory, the students in a particular college, the books in a library, are all finite populations. A population which is not finite is said to be infinite. 'All possible tosses of a coin' is an infinite population for, although there may be a first member, there can never be a last.

We have so far in this book been dealing with descriptive statistics. That is, that branch of the subject which is concerned with the presentation of data in tables and diagrams, and with the calculation of quantities such as the arithmetic mean and standard deviation which summarize certain aspects of the data. Since the characteristics of a population are likely to differ to some extent from those of a sample, it is usual to use different words to describe these characteristics. We speak of population characteristics as population parameters and sample characteristics as sample statistics.

If we are going to use the sample to obtain information about the population from which the sample was drawn then, if this information is to be useful, the sample should be representative of its parent population. Any selection will not provide a satisfactory sample. Some may even be misleading. If we wished to estimate the mean weight of boys in the senior classes of a school by selecting a sample of fifteen boys and finding their mean weight, the first-grade football team would not be a satisfactory sample for this purpose. This would be a biased sample.

Such special samples must not be used if we are to obtain reliable information about populations. Many sampling techniques have been developed to enable the statistician to obtain samples which yield reliable information about the population. One of these is called simple random sampling. Here samples are selected in some purely arbitrary fashion. There is no deliberate choice on the part of the observer, no attempt to select 'typical' individuals. Some chance procedure, such as selection by lot, is used to produce what are called random samples. Macquarie University might produce a sample which is not representative of Australian university students. Reliable information about Australian university students could however be obtained by selecting students at random from *all* universities. Another method which could be used is to select a random sample from every university in Australia and combine the results in a way which makes appropriate allowances for the various sizes of the different universities. We will not be concerned in this book with the theory and practice of sampling but it is important in statistical work to understand the concept of random sampling.

Statistical regularity

Individuals in a population vary. Australian adult males have varying heights; families in NSW have varying numbers of children; the number of motor vehicle accidents varies from year to year; seat belts for motor cars have varying breaking strengths. However, populations have a stability not possessed by individuals. The proportion of adult males over 190 cm. in height in the whole population from year to year is remarkably constant, as is the proportion of childless families.

These stable properties of populations are reflected in sufficiently large samples drawn from the population. The sex of an individual unborn child is unpredictable; yet the proportion of male births in the whole population of births or in large samples from it, is remarkably constant.

Most people, perhaps unconsciously, are aware of this phenomenon of statistical regularity. This awareness prompts their casual references to the 'law of averages'. Insurance offices depend for their success on the fact that death, accidents and sickness cannot be easily predicted in individuals but show remarkable regularity in large groups. No one knows when a coin is tossed whether it will be a head or a tail, but in a large number of tosses everyone knows that roughly one half will be heads.

Probability

We can describe this statistical regularity in the language of statistics. Let us assume that the observations of a sample are set out in the form of a frequency table; for example, the number of throws of a six-sided die which produce a one, a two, a three etc. The relative frequencies are the ratios of the frequency of a one, a two, a three and so on to the total frequency. The principle of statistical regularity may be stated in the following way:

Provided the samples are random, the relative frequency of occurrence of a particular event (a one, a two, a three, etc.) in the sample approaches nearer and nearer to the value in the population as the sample size is increased.

The population value which this relative frequency approaches is called the probability of this particular event. Thus the probability of getting a one, with an unbiased six-faced die is 1/6. Since the frequency of any value cannot be less than 0 and cannot exceed the total frequency, the probability of a particular event cannot be less than zero nor greater than one.

Probability of mutually exclusive events: either A or B

Mutually exclusive events are ones which cannot occur together in the same experiment. For example, getting a one and getting a two in a single throw of a die are mutually exclusive events. If the occurrence of either one or other of two mutually exclusive events is considered satisfactory, then the probability of a satisfactory outcome is the sum of the probabilities of the two mutually exclusive events. The probability of getting a one is 1/6; the probability of getting a two is 1/6. Hence the probability of getting either a one or a two is 1/6 + 1/6 or 1/3.

Rule: To find the probability that any one of a number of mutually exclusive events will occur, *add* the individual probabilities.

Probability of two independent events both occurring: A and B

Two events are independent if the happening of one can have no effect on the other. Throwing a six with a die can have no effect on the next throw; they are independent. However, if there are three black and three white balls in a bag and we draw balls from the bag without re-placement, the second draw is not independent of the first because our chances of getting a white ball on the second draw depend on the result of the first draw.

Rule: The probability that *both* of two independent events will occur is the *product* of their individual probabilities.

For example, the probability of throwing two sixes in two throws of a die is 1/6 × 1/6 or 1/36.

If the probability that a male, selected at random from a very large population, is over 190 cm. in height is 1/10, the probability of getting two men over 190 cm. in two random independent selections from this large population is 1/100. The probability of getting six men of this height or more in six random independent selections would be $1/10^6$ or one in a million.

Example 5.1

Let us consider the simple experiment of tossing for tails with an un-biased coin. If the experiment consists of one toss only then this is a sample from a population with the probability distribution shown in *Fig. 5.1.*

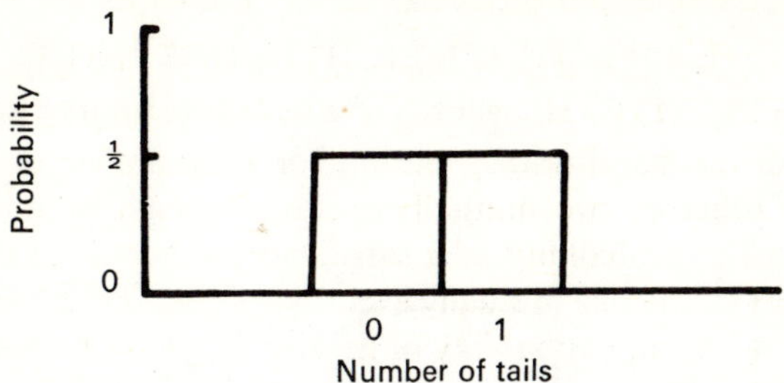

Fig. 5.1

Probability distribution for number of tails with one toss of an unbiased coin

If the rules of the experiment are changed and we are allowed two tosses, then the possible outcomes of the experiment are

$$TT, \quad TH, \quad HT \quad and \quad HH$$

where T stands for tails and H for heads.

Each of these outcomes is equally likely to occur (with an unbiased coin), that is each outcome would be expected in roughly one quarter of the experiments. The number of tails would therefore be two in about one quarter of the experiments, one in about one half of the experiments and nil in about one quarter. In other words, the infinite population of two throws of an unbiased coin has the probability distribution shown in *Fig. 5.2*.

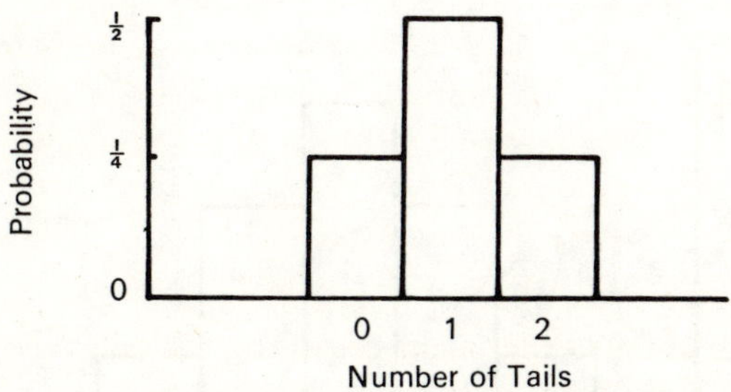

Fig. 5.2

Probability distribution for number of tails with two tosses of an unbiased coin

With three tosses of a coin the equally likely outcomes are

TTT, TTH, THT, THH, HTT, HTH, HHT, HHH

or T^3, $3T^2H$, $3TH^2$, H^3 where T^3 stands for 3 tails etc.

The population probability distribution for the number of tails is shown in *Fig. 5.3*.

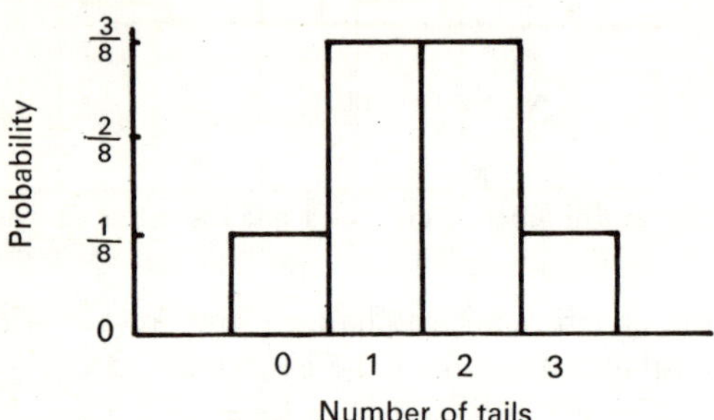

Fig. 5.3

Probability distribution for number of tails with three tosses of an unbiased coin

With a four, five, ten, one hundred etc. coin tossing experiment we can determine in the same way the probability of the various possible outcomes.

The population probability distributions in these cases are all shown in *Fig. 5.4*.

FOUR TOSSES

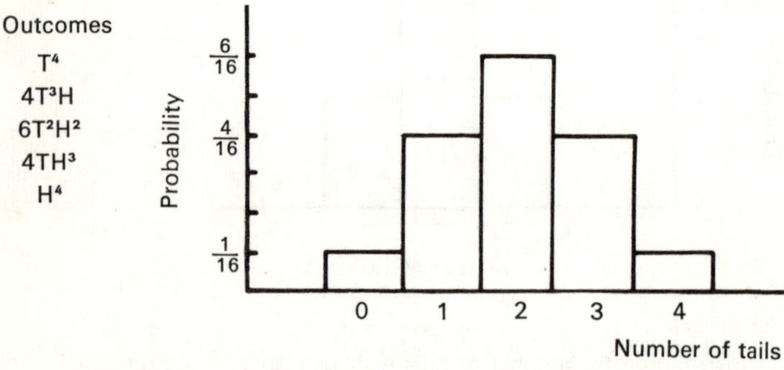

FIVE TOSSES

Outcomes

T^5

$5T^4H$

$10T^3H^2$

$10T^2H^3$

$5TH^4$

H^5

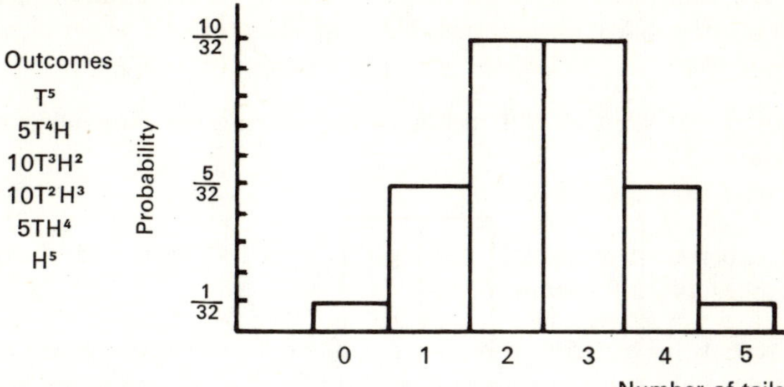

TEN TOSSES

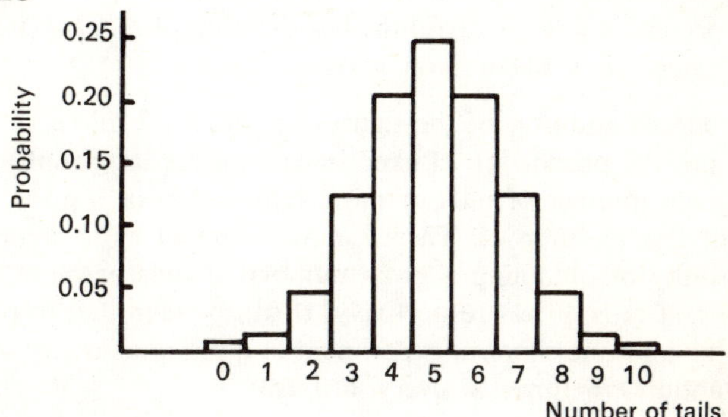

ONE HUNDRED TOSSES

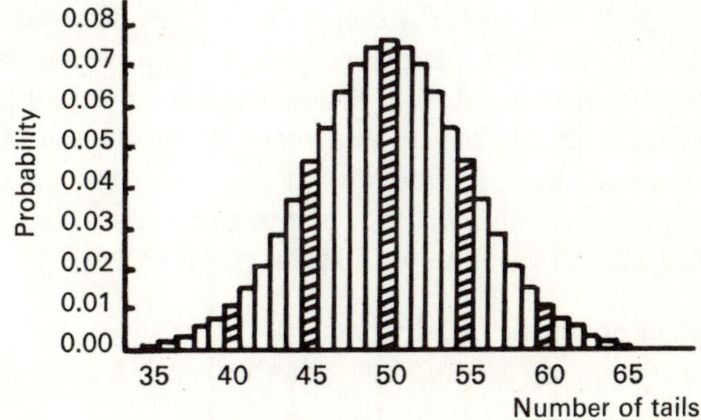

Fig. 5.4

Probability distributions for number of tails with four, five, ten and one hundred tosses of an unbiased coin

Note: The total area under all these probability distributions equals the total probability of all possible outcomes of the experiment, namely, unity.

Let us repeat what all this means by considering the experiment of 100 tosses:
1. We have an unbiased coin.

2. Our experiment consists of tossing this coin 100 times and noting the number of tails obtained.

3. We repeat this experiment millions of times. The relative number of times we would obtain various numbers of tails can be worked out and are shown in the second column of *Table 5.1*. It is these figures which are represented graphically in the last diagram of *Fig. 5.4*. It shows how common or how uncommon various scores are.

By simple addition of the figures in column 2 of *Table 5.1* we can work out the proportion of experiments (in the long run) which result in a given number of tails or more (column 3) or a given number of tails or less (column 4). These columns enable us to write down the probability of obtaining a given number of tails or more, or a given number of tails or less respectively. It can be seen that to obtain 65 or more tails in one random experiment is quite unusual. It would occur only about 176 times in every 100,000.

Is a given coin unbiased?

All we have dealt with so far is theory, based on the assumption of an *unbiased* coin. In practice we generally do not know whether a coin is biased or not; that is, we do not know whether in the long run it produces more tails than heads or vice versa, or heads and tails in equal numbers. We may wish to find out, from some experiments we have carried out with this coin, whether or not there is strong evidence that it is biased. *Table 5.1* can be used for this purpose.

We now set down the usual steps:
1. We have a coin; we do not know whether it is biased or not.

2. One possible experiment consists of tossing the coin 100 times and noting the number of tails obtained. Let us assume for purposes of explanation that we do this once and that the number of tails obtained

Table 5.1

Theoretical outcome if the experiment of tossing an unbiased coin 100 times
is repeated indefinitely

Number of tails obtained (x)	Number of experiments per 100,000 which result in		
	exactly x tails	x or more tails	x or less tails
100	0*	0*	100,000
99	0*	0*	100,000
.	.	.	.
.	.	.	.
70	2	4	99,998
69	5	9	99,996
68	11	20	99,991
67	24	44	99,980
66	45	89	99,956
65	87	176	99,911
.	.	.	.
.	.	.	.
60	1,084	2,844	98,240
.	.	.	.
.	.	.	.
55	4,847	18,410	86,437
54	5,796	24,206	81,590
53	6,659	30,865	75,794
52	7,353	38,218	69,135
51	7,803	46,021	61,782
50	7,958	53,979	53,979
49	7,803	61,782	46,021
48	7,353	69,135	38,218
47	6,659	75,794	30,865
46	5,796	81,590	24,206
45	4,847	86,437	18,410
.	.	.	.
.	.	.	.
40	1,084	98,240	2,844
.	.	.	.
.	.	.	.
35	87	99,911	176
34	45	99,956	89
33	24	99,980	44
32	11	99,991	20
31	5	99,996	9
30	2	99,998	4
.	.	.	.
.	.	.	.
1	0*	100,000	0*
0	0*	100,000	0*
Total 100,000			

* Not actually zero, but extremely small

is sixty-five. This is the result of our experiment and we proceed to analyse it.

3. We make the assumption that the coin was unbiased. Table 5.1 tells us that, *with an unbiased coin,* one experiment producing 65 or more tails would only occur in 176 experiments out of every 100,000 carried out, that is about once in every 600.

4. Our experiment which produced 65 tails is therefore, *with an unbiased coin,* a most infrequent occurrence. Hence we must conclude:
Either (i) Our coin is unbiased and we have performed a near miracle;
or   (ii) Our assumption is wrong and the coin is in fact biased.
   The alternative (ii) seems the more reasonable conclusion.

Applications

The reader might very well wonder what application such an example has to experimental work in the fields of education, psychology, and the sciences. It has in fact a wide range of applications. I shall mention three only. These are experiments which from a statistical point of view are identical with the example we have studied. The background is very different in each case, but the statistical analysis is identical.

1. Acoustics: A few years ago I was keenly interested in 'hi-fi' equipment and an acoustics engineer told me he could effect a considerable improvement by an addition to the speaker system. To convince me he made the addition and kept pointing out what an improvement it had made. Frankly, I could not detect any difference and began to wonder whether my perception was at fault.

   I therefore asked him to help me perform an experiment. We placed a switch in the circuit by which the addition to the system could be included or excluded at will. I operated the switch and he was to tell me whether the 'addition' was operating or not. He was given 100 trials and I recorded his decisions.

   The analysis proceeds exactly as in the previous example. Assume that he cannot detect any difference and that his decisions are made at random (like the fall of an unbiased coin). If his decisions were correct in a sufficiently large number of cases (as judged by the figures in *Table 5.1*) then we should probably conclude that our assumption was wrong and that he can in reality detect differences.

   In actual fact, no calculations were necessary as he was wrong more often than he was right.

2. Education: Students are often set tests consisting of 100 questions of the 'true or false' variety. It is claimed that students knowing nothing about the subject and selecting answers purely at random (like the fall of an unbiased coin) may achieve reasonable results. This assumption also may be tested as in the example.

3. Psychology: McCollough and Van Atta[1], describe an experiment in which they are studying the behaviour of rats in a T maze.

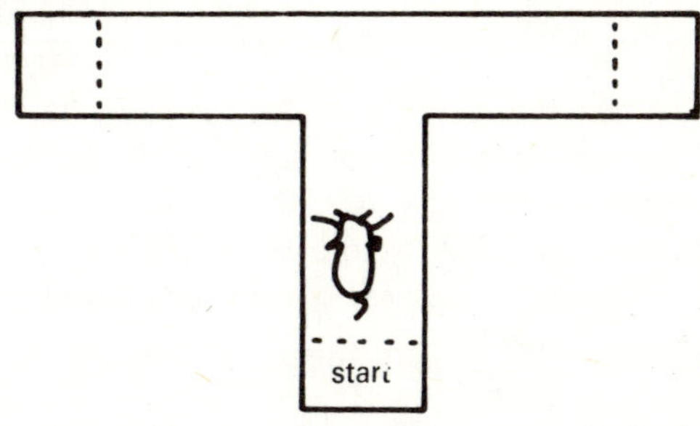

Fig. 5.5

A rat is released from the start box and allowed to run into either of two end boxes at the ends of the right and left arms of the T. Neither end box contains food or water. On reaching an end box he is placed for a second time in the starting position. It is noted whether he runs to the same box as on the first trial (which we shall call a 'repeat') or to the other box (which we shall call a 'reverse'). We are interested to see if there is any curiosity motive or exploratory drive, or whether they generally follow the same path, and hence are creatures of habit.

If we assume that our rats are of the type that choose to repeat or reverse purely at random then repeats and reverses are equally likely to occur. An experiment with 100 such rats is statistically the same as tossing an unbiased coin 100 times. We could change the words 'number of tails' in our example to 'number of repeats' and our analysis would apply to this experiment.

If 65 or more of the 100 rats chose to repeat then we could conclude:
Either (i) the rats act purely at random and we have performed a near
    miracle (a 1 in 600 chance)
or (ii) our assumption is wrong and the rats are creatures of habit.

The alternative (ii) seems the more reasonable conclusion.

[1] C. McCollough and L. Van Atta, *Statistical Concepts,* McGraw Hill, New York, 1963

Determining probabilities from the distribution of Fig. 5.4

The method of determining probabilities associated with the 100 coin tossing experiment can best be seen by considering a few examples. We shall refer to the last graph in *Fig. 5.4* and to *Table 5.1* in which the actual numerical values are recorded.

From the second column in *Table 5.1*, 45 tails would be the outcome in about 4,847 experiments out of every 100,000. Hence the probability of obtaining a value of 45 tails in one experiment is $\dfrac{4,847}{100,000}$ or 0.04847.

This is the area of the bar standing above the value 45 in *Fig. 5.4*.

Again from the second column of *Table 5.1* it can be seen that the number of experiments in every 100,000 in which the outcome is 46, 47 or 48 tails would be 5,796 + 6,659 + 7,353 or 19,808. Hence the probability of obtaining in one experiment any number of tails from 46 to 48 inclusive is $\dfrac{19,808}{100,000}$ or 0.19808. This is the area in *Fig. 5.4* of the three bars standing above the values 46, 47 and 48. We can obtain this same value most easily by using the last column in *Table 5.1*. The chance of getting 48 tails *or less* is 38,218 in 100,000 and the chance of 45 *or less* is 18,410 in 100,000. The difference, which must be the chance of 46, 47 or 48 tails, equals $\dfrac{38,218}{100,000}$ minus $\dfrac{18,410}{100,000}$ or 0.19808 as before.

The use of the cumulative figures in the last column of *Table 5.1* avoids the need for long additions.

The chance in one of these experiments of obtaining any number of tails from 33 to 66 inclusive equals $\dfrac{99,956}{100,000}$ minus $\dfrac{20}{100,000}$ or 0.99936.

The chance of obtaining *less than* 55 tails equals the area of all the bars standing *to the left* of the bar above the value 55. From *Table 5.1* this chance is 0.81590.

The probability of obtaining less than 55 tails in each of four independent experiments, from the previous paragraph and the product law for probabilities, equals $(0.8159)^4$ or 0.4431.

Exercises

5.1 A family consists of 4 children. If the probability that each child is a male is 1/2 what is the probability that
a. the family consists of 1 boy and 3 girls
b. the family consists of 2 boys and 2 girls
c. the two oldest will be boys and two youngest girls
d. the eldest will be a boy and the rest girls?

5.2 A and B are two individuals whose chances of dying in a year are .04 and .03 respectively. What is the chance that during the year
a. A will die and B live
b. A and B will both die
c. A and B will both live
d. At least one of A and B will die?

5.3 A psychology lecturer wishes to find out about the television watching habits of the community. To do so he asks his class to complete a questionnaire. Comment on the reliability of his results.

5.4 An experiment consists of tossing a coin four times and noting the number of tails. The experiment was carried out 20 times and the frequencies of occurrence of 0, 1, 2, 3 and 4 tails were 2, 5, 6, 5 and 2 respectively. Use the formulae on pages 33 and 47 to calculate the mean and standard deviation for the number of tails obtained.

5.5 Repeat question *5.4* given that the experiment now consists of tossing a coin five times and that the frequencies of obtaining 0, 1, 2, 3, 4 and 5 were 0, 7, 9, 11, 3 and 2 respectively.

5.6 In tossing an unbiased coin 100 times, find from *Table 5.1* the chance of obtaining
a. 48 to 54 tails
b. more than 46 tails
c. 52 tails or less.

6

The Normal Distribution

It can easily be seen that as the number of tosses of a coin (or the number of rats used) in the experiment described in Chapter 5 is increased, the probability distribution approaches nearer and nearer to a smooth curve. Very often in practice, when we are dealing with populations with a large number of observations, the frequency of observations of various sizes seems to follow quite closely the pattern of this particular bell-shaped curve. The IQs, the heights, the weights of the adult male population, yield per acre of varieties of wheat and so on seem to approach this pattern. It (or approximations to it), occurs so often in practice and it is so useful in statistical theory, that it was recognised early and called the normal distribution.

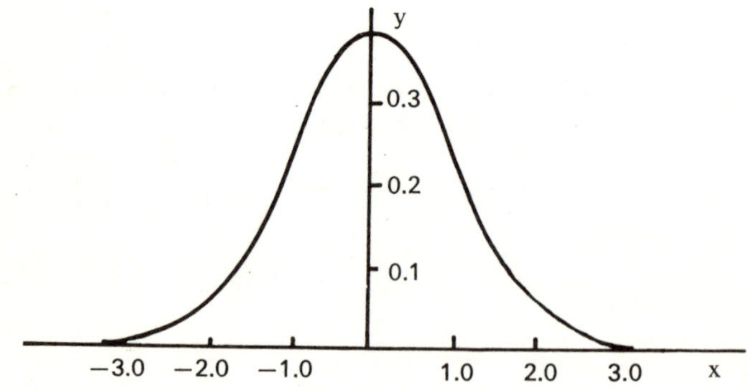

Fig. 6.1
The standard normal distribution

The normal probability curve

If we were to carry out a very large number of these 100 coin tossing experiments with an unbiased coin, the formula on page 84 shows that the mean of the population is 50 and the standard deviation five. If we were to make our measurements from 50 (i.e. the mean) and change the units of deviation to 5 (i.e. units of standard deviation) then the normal probability curve to which this population approximates would be as shown in *Fig. 6.1.*

The units are standard scores which we discussed in Chapter 4. The total area under the curve still equals unity and as the horizontal units have been reduced fivefold the vertical units have been multiplied by five.

Extensive tabulations have been made of the area under the curve between the *y*-axis and any other vertical line. One such tabulation is given in the *Appendix* on page 256. From this table we deduce

1. The area to the right of the y axis $= 0.5000$
2. The area between the mean and $+1$ standard deviation $= .3413$
3. The area between the mean and $+2$ standard deviations $= .4772$
4. The area between the mean and $+2.6$ standard deviations $= .4953$
5. The area between $+1$ SD and $+2$ SD $= .1359$
6. The area between -1 SD and $+1$ SD $= .6826$
7. The area between -2 SD and $+2$ SD $= .9544$
8. The area between -1 SD and $+2$ SD $= .8185$

The student must develop speed and accuracy in calculating all possible areas under the normal curve by means of this table as these areas are required in the solution of many problems.

Notation

We shall use μ (the Greek letter for *m* called *mu*) and σ (the Greek letter for *s* called *sigma*) to denote the mean and standard deviation of a population. For a sample, we shall use $\bar{x}$ and s to denote the mean and the standard deviation.

Using tables of the normal distribution

1. If we are told that the height of the Australian male adult population has a mean of 175 cm and a standard deviation of 10 cm how many men out of a typical group of 10,000 would we expect to have heights exceeding (*a*) $177\frac{1}{2}$ cm   (*b*) 180 cm   (*c*) $182\frac{1}{2}$ cm   (*d*) 185 cm   (*e*) $187\frac{1}{2}$ cm ? How many would have heights between $166\frac{1}{4}$ cm and $181\frac{1}{4}$ cm ?

From the table in the *Appendix* (page 256) the proportions of a normal population which exceed the mean by $\frac{1}{4}\sigma$, $\frac{1}{2}\sigma$, $\frac{3}{4}\sigma$, σ, $1\frac{1}{4}\sigma$, are .4013, .3085, .2266, .1587, .1056. Hence if we can assume that the heights follow a normal distribution with $\mu = 175$ cm and $\sigma = 10$ cm then the answers to (*a*) — (*e*) are 4013, 3085, 2266, 1587 and 1056 men.

The number between $166\frac{1}{4}$ cm and $181\frac{1}{4}$ cm in height

$\quad$ = the number between $8\frac{3}{4}$ cm less than and $6\frac{1}{4}$ cm greater than the mean

$\quad$ = the number between $.875\sigma$ less than and $.625\sigma$ greater than the mean

$\quad$ = 10,000 (.3092 + .2340)

$\quad$ = 10,000 × .5432

$\quad$ = 5,432 men.

2. In a statewide examination my marks, as well as the mean and standard deviation of all the marks, are set out in the following table:

	Mean	SD	My mark
Latin	53	12	81
French	61	8	81

What proportion of the candidates were above me?

My standard scores (i.e. number of standard deviations from the mean) were

$$\text{Latin} \qquad \frac{81 - 53}{12} = +2.33$$

$$\text{French} \qquad \frac{81 - 61}{8} = +2.5$$

The area under the normal curve above 2.33σ

$$= .5000 - .4901 = .0099$$

This is almost .01 or 1% of the total area. For french the proportion is .5000 — .4938 = .0062 or about 0.6%.

Thus assuming the marks were normally distributed the percentages above me can be taken to be

$$\text{Latin } 1.0\%; \text{ French } 0.6\%.$$

3. If IQs are normally distributed with $\mu = 100$ and $\sigma = 10$, what is the probability that a randomly selected individual has an IQ between 90 and 120?

This probability equals the probability that his standard score lies between —1 and +2. This equals .4772 + .3413 = .8185.

Samples from a normal population

Our usual purpose in selecting samples from a population is to find out information about the population itself. If I wish to guess the average height of the Australian adult male population I could be a long way out if I selected one man at random, measured his height and used this figure as my guess. I would be better advised to select 100 men at random and find their average height. This should be nearer the mark. If I selected 10,000 and found their average height I would do better still. The size of the sample is obviously important.

There is an important theorem which says that if a population is distributed normally with mean μ and standard deviation σ, then if we take samples of size n from that population, the sample means $\bar{x}$ will

be normally distributed with mean μ and SD $= \dfrac{\sigma}{\sqrt{n}}$

For example, assume the height of the Australian adult male is normally distributed with mean 175 cm and SD = 10 cm. If I select 100 men and measure their heights and calculate the mean, you select another 100 and calculate their mean height, and thousands of other people carry out the same experiment, we will find that the means we all get will vary, but the mean of these means will be about 175 cm

and the SD of the means will be about $\dfrac{10}{\sqrt{100}}$ or 1 cm.

If each of us had selected samples of 10,000 to work out our mean height, then we would have found these mean heights to vary around

175 cm with a SD of $\dfrac{10}{\sqrt{10,000}}$ or 1/10 cm. In short, the larger the number in the sample the less its mean will vary from the population mean.

Central Limit Theorem

This leads us to one of the most remarkable theorems in mathematical literature, the Central Limit Theorem. It is this theorem which explains why the normal distribution is so important in statistics.

If we select samples of size n from a population in which the quantity x varies with mean μ and SD σ but is *not* normally distributed, then the mean of the samples $\overline{x}$ varies *approximately normally* with mean μ and standard deviation $\dfrac{\sigma}{\sqrt{n}}$, the approximation becoming increasingly good as n increases.

This indicates that, if we are working with the mean of reasonably large samples, we can use normal distribution tables even if the underlying population is not really normally distributed.

This remarkable theorem is probably best illustrated by considering the infinite population of throws of an unbiased six-sided die. The population probability distribution is far from normal, as *Fig. 6.2* shows. Each score is equally likely.

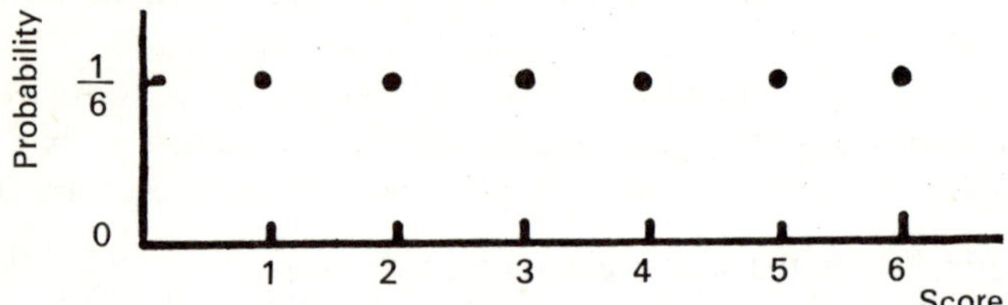

Fig. 6.2

Probability distribution for the score with one throw of an unbiased six-sided die

If we select a sample of *one* throw from this population, then again all values are equally likely and the probability distribution is as in *Fig. 6.2*.

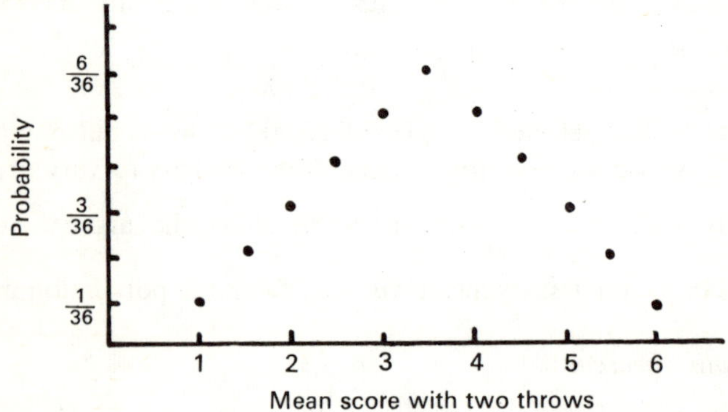

Fig. 6.3

Probability distribution for the mean scores with two throws of an unbiased six-sided die

If however we select a sample of *two* throws from this non-normal population and take the mean we are far more likely to obtain values of (say) $2\frac{1}{2}$, 3, $3\frac{1}{2}$, 4, than 1 or 6. The probability distribution of the mean of samples of two can easily be shown to be as in *Fig. 6.3*.

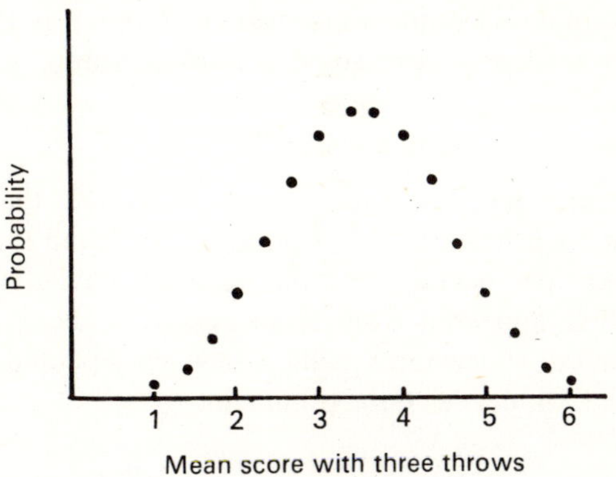

Fig. 6.4

Probability distribution for the mean score with three throws of an unbiased six-sided die

For the mean of samples of *three* we obtain the distribution of *Fig. 6.4*.

The approach to normality is even now becoming apparent. For the larger samples the distribution of sample mean may be taken as normal in spite of the most non-normal nature of the population.

Example 6.1

We are given that the distribution of IQs in the population has $\mu = 100$ and $\sigma = 10$. A sample of 25 individuals was selected and their average IQ was found to be 94. Comment.

Solution

If IQs are known to be normally distributed, no approximation is involved. With a sample of 25 however, the mean of the sample is approximately normally distributed whatever the population IQ distribution.

If we select many random samples of 25 from the population their means would vary about a mean of 100 with SD of $\dfrac{10}{\sqrt{25}}$ or 2.

We have thus selected one sample with a mean IQ of 94 from a population of samples whose mean varies with mean 100 and SD of 2. Hence the mean of our sample varies from the mean of the population of samples by 3 standard deviations. The chance of a variation of this amount or greater is .0013 , that is, there is only a chance of about 1 in 1,000 of obtaining a sample which varies as much as this from the population. We evidently have some abnormal group.

Example 6.2

A scientific farmer has over many years measured the yield of a certain crop on each hectare he has planted. He found that the mean yield was 82 kg per hectare and the standard deviation 4 kg per hectare. In what appeared from other results to be a quite usual season, he planted 16 hectares with a new variety and obtained a mean yield of 85 kg per hectare. Comment.

Solution
For purposes of analysis we first make the assumption that the yield for this variety is in reality no different from that of previous varieties. That is, we assume that the result from the 16 hectares is a random sample from the same population as in the past and that the better result is just a chance fluctuation.

With samples of 16 hectares we may assume from the Central Limit Theorem that the yield of the 16 hectares is normally distributed (whatever the parent distribution) with a mean of 82 kg and a standard deviation of

$$\frac{4}{\sqrt{16}} \text{ or } 1 \text{ kg.}$$

This particular sample has a mean yield of 85 kg. That is, it differs from the mean by three standard deviations. From normal distribution tables, a difference of this amount or greater would only occur by chance 13 times in 10,000. This is a most unlikely result. If the season has been a quite usual one and the soil and treatments have not been varied he could be reasonably satisfied that the new variety has a higher yield.

Exercises

6.1 The height of male university students is normally distributed with mean 175 cm and standard deviation 9 cm. Calculate the probability that the height of a student selected at random is
a. Less than 163 cm
b. Between 163 cm and 178 cm.

6.2 In a large city the IQ of students on leaving school is normally distributed with a mean of 100 and a standard deviation of 10. The education authorities divide the students into three groups:
a. Those with an IQ of less than 98 they consider should receive no further formal education
b. Those with an IQ between 98 and 110 they recommend for a technical education
c. Those with an IQ above 110 they recommend for a university education.
   There are 10,000 students altogether in this particular year.
How many would be in each of the groups?

6.3 In a state-wide public examination A passes are awarded in a certain subject to 20% of students. If the mean mark was 55 and the standard deviation 20 what would the minimum A mark be?
(You may assume the marks were normally distributed.)

6.4 A manufacturer of 10 cm bolts has told his foreman that not more than $1\frac{1}{2}\%$ of the output of bolts is to be rejected because of departure from the length specification. The foreman knows from past experience that the mean length of bolts produced by the machine is 10 cm and the standard deviation is 0.2 cm. What maximum and minimum lengths should he specify so as to just comply with his employer's instructions?
(You may assume that the length of the bolts is normally distributed.)

6.5 The contents of a bottle of lotion is claimed to average 80 cc with a standard deviation of 4 cc. What is the chance that
a. The contents of one bottle of lotion lies between 79 cc and 81 cc?
b. The average contents of 100 bottles will be between 79 cc and 81 cc?

6.6 If x is normally distributed with mean 30 and standard deviation 8 what is the probability that the mean of a sample of 16 will
a. Be less than 32   *b.* Exceed 36   *c.* Exceed 28   *d.* Be less than 25
e. Be between 33 and 34?

7

Testing Hypotheses about Populations: Normal Populations

It is usual to use the results of a sample to test some hypothesis which may have been made about the population from which it was drawn. We shall illustrate the approach of the statistician by considering some examples.

Testing means — population variances known

Example 7.1

A biologist is studying the growth rate of rats. He has found from experiments with a large number of rats on a given diet, that the increase in weight in the first six months of life is normally distributed with mean 262 grams and standard deviation 22 grams. He is given a rat of a different species and subjects it to the same treatment. Its mean increase in weight is 300 grams. He knows nothing about the growth rate of this new species and would like to know whether it is just the same as those with which he is familiar or whether it is a rapidly growing species. How should he tackle the problem?

Solution

1. Hypothesis Let us make the assumption that there is no difference in the growth rates of the new species and of the old — that is, that the new rat is from a species which also has a growth rate normally distributed with mean 262 grams and standard deviation 22 grams. This is called the *hypothesis*. (Sometimes this assumption of 'no difference' is called a 'null hypothesis'.)

2. Deduction If the new rat is from such a population then his growth is 38 grams above the mean, that is 1.73 standard deviations above the mean. From normal tables we see that an increase of this amount or greater would only occur *by chance* about once in 25 times.

3. Conclusion There are two possible conclusions:

Either (*i*) our hypothesis is correct — the new species has the same growth rate as the old — but we have encountered by chance a quite unusual rat of his species

or   (*ii*) the hypothesis on which we made our calculation is *not* correct and hence the new species has a greater growth.

4. Comment The alternative (*ii*) seems the more reasonable conclusion. We could of course be wrong in coming to this conclusion if we happen to have encountered a quite unusual rat. In this case there is a once in twenty-five chance of being wrong.

Example 7.2

The biologist wishes to be more sure of his conclusion. He therefore repeats the experiment with 10 rats of the new species and finds the mean growth 286 grams. What conclusions can he now draw?

Solution

1. Hypothesis Assume that the new species has the same growth rate as the old

$$\text{i.e. } \mu = 262 \text{ grams}$$
$$\sigma = 22 \text{ grams}$$

2. Deduction On this hypothesis the mean growth of a sample of ten from this population would be normally distributed with mean 262

grams and standard deviation $\dfrac{22}{\sqrt{10}}$ or 7 grams. Our particular sample of

10 has a mean of 286, that is $\dfrac{24}{7}$ or 3.43 standard deviations above 262

(the mean of a large number of such samples). From normal tables, a growth of this amount or greater would occur by chance less than once in 1,000 times.

3. Conclusion There are two possible conclusions:

Either (*i*) our hypothesis is correct — the new species has the same growth as the old — but we have encountered by chance a most unlikely set of 10 rats of this new species

or   (*ii*) our hypothesis is *not* correct and hence the new species has a greater growth.

4. Comment The alternative (*ii*) seems the more reasonable conclusion. We could, of course, be wrong in coming to this conclusion if we happen to have encountered a most unusual set of 10 rats. In this case our chance of being wrong is less than 1 in 1,000.

Example 7.3

A professor of psychology wants to know whether his students this year obtained significantly poorer results in statistics than students have generally obtained in the past. Experience has shown that the marks have been normally distributed with a mean of 50 and a standard deviation of 8. There are this year 400 psychology students with a mean mark of 48.4.

Solution

1. Hypothesis Assume that there is no real difference between this year's psychology students and students in the past; that is, that the psychology students are a random sample of 400 students from a population which has a mean mark of 50 and a standard deviation of 8.

2. Deduction On this hypothesis the means of samples of 400 would be normally distributed with mean 50 and standard deviation $\dfrac{8}{\sqrt{400}}$, or 0.4. The value we have obtained for the 400 psychology students is 48.4 i.e. $\dfrac{1.6}{0.4}$ or 4 standard deviations below the mean. Normal tables show that a mean mark this amount or more below 50 would occur by chance far less than once in 1,000 times.

3. Conclusion There are two possible conclusions:

Either (*i*) our hypothesis is correct — that these psychology students are *not* poorer than former students in statistics — and we have by a remarkable chance (less than 1 in 1,000) come across an unusual collection of 400 students

or (*ii*) our hypothesis is *not* correct and that these psychology students are poorer than former students in statistics.

4. Comment The alternative (*ii*) seems by far the more reasonable assumption. We could, of course, be wrong in this assumption but the chances of this are less than 1 in 1,000. It seems that we can be almost certain from this result that his psychology students this year are poorer than the average students in statistics.

Testing the difference between two means: population variances known

Often we need to compare the means of two populations which we know have different backgrounds but we do not know whether these different backgrounds have any effect on the quantity we are measuring. For example, are psychology students better than education students in statistics? Is one drug better than another? Is one fertilizer better than another? Is one diet better than another?

The comparison usually has to be made by comparing the mean of a sample from one population with the mean of a sample from the other. We make the null hypothesis that there is no difference between the means of the two populations. They are both μ (say). The standard deviations of the populations are known and are σ_1 and σ_2 respectively. If the first sample is of size n_1 its mean will be distributed with mean μ and with standard deviation $\dfrac{\sigma_1}{\sqrt{n_1}}$. If the second sample is of size n_2, its mean will also be distributed with mean μ but with standard deviation $\dfrac{\sigma_2}{\sqrt{n_2}}$. We are interested in the *difference* between the two means. It can be shown that if these are independent random samples the *difference* between the two means will be distributed with zero mean and with standard deviation equal to

$$\sqrt{\frac{\sigma_1{}^2}{n_1} + \frac{\sigma_2{}^2}{n_2}}$$

That is to say that if we were to select two independent samples, one from each of these populations, and were to calculate the difference between their means and repeat the performance very many times we would find the mean of the figures obtained would be zero and the SD equal to

$$\sqrt{\frac{\sigma_1{}^2}{n_1} + \frac{\sigma_2{}^2}{n_2}}$$

Hence we can tell, according to our hypothesis, how unusual it would be to get the *actual difference* we have obtained between these two particular samples. This should enable us to tell whether they are likely to have come from populations with the same mean.

If our hypothetical population is not normal, Central Limit Theorem tells us we can still use this method if our samples are large enough.

Example 7.4

The 400 students majoring in psychology in a particular year obtained a mean mark of 48.4 in statistics. For the 300 students majoring in the same year in education the mean mark was 49. 8. Experience over a long period has shown that statistics marks are normally distributed with a standard deviation of 8. Can we presume that psychology students and education students performed equally well in statistics?

Solution

1. Hypothesis Assume the null hypothesis that there is no real difference in their performance and that the result we have obtained is simply a chance variation.

2. Deduction On this hypothesis, we have two random samples from the same normal population. The difference of their means is therefore normally distributed with mean zero and standard deviation

$$\sqrt{\frac{64}{400} + \frac{64}{300}} = 0.61.$$

The difference we have obtained between the mean marks is 1.4 which differs from zero by 2.30 standard deviations. From normal tables this difference or an even greater one is only likely to occur about once in fifty times.

3. Conclusion There are two possible conclusions:

Either (*i*) our hypothesis is correct — there is no real difference in their performance in statistics — and we have obtained one of those unusual groups of students which one expects to encounter by chance once in 50 times

or   (*ii*) our hypothesis is *not* correct and there is a real difference in performance.

4. Comment The alternative (*ii*) seems the more reasonable conclusion. We could of course be wrong in this conclusion; in fact, we would expect to be wrong about once in 50 times. This, however, is a rare event and it is unlikely that this is the one year in fifty in which it occurs. We can accept that there is a difference in performance with reasonable confidence.

Summary

From these examples it can be seen that the aim of hypothesis testing is to determine, *from the information we have in a sample*, whether the population from which it is drawn is likely to have certain specified characteristics. In some of the examples which we have considered so far we were concerned with determining from the sample whether the population was likely to have had a certain mean. The same logical approach may be made to determine whether the population is likely to have had a given median, or a given proportion, or a given standard deviation and so on. The logic is the same but the statistical analysis different. We shall deal with some of these other cases of hypothesis testing later.

The need for hypothesis testing arises because of the variations which occur in nature. A sample is generally not a perfect miniature of its population. Some chance variation is expected. For example, the mean of a sample is not likely to be exactly equal to the mean of the population. Large differences, however, rarely occur; if the difference is large enough we may very well be suspicious that the sample in question was not in fact a random sample from the given population. To determine what is a large difference or an unusual difference between sample and population for the particular characteristic we are studying (e.g. mean, median, proportion etc.) we need to know how frequently various sample values of this characteristic are likely to be obtained in samples from the population. Statisticians have prepared tables setting out the probability of obtaining certain values in a sample from given populations for all the types of problem with which we shall be concerned. If the probability that the sample value could have occurred by chance in random samples selected from the population is very small, we may then, with reasonable confidence, conclude (knowing that occasionally we may be wrong) that the sample has not been drawn from the given population. The smaller the probability the more confident we are in our conclusion.

We have, however, to decide just how unlikely a sample has to be before we reject the hypothesis that it came from the specified population and presume that some cause other than chance has produced the large deviation. A typical limit is a deviation which would only be exceeded by chance once in 20 times. Deviations greater than this are then said to indicate 'significant' differences between the sample and the population in the characteristic being studied. We must remember that, if we use this particular limit, the conclusion which we draw will be wrong once in 20 times. We can of course reduce the chance of this error by changing our limit of significance to once in 100 times (say), but the deviation then required to detect a significant difference becomes so large that we may well not detect a difference when one really does exist.

The once in 20 times and once in 100 times criteria, known respectively as 5% and 1% significance levels, are most commonly used in practice. The choice of significance level in practice depends on the nature of the particular problem under study.

We can see from the examples dealt with so far that the steps in hypothesis testing are quite standard.

1. We set up a hypothesis concerning the population.

2. We determine from tables the chance that the particular sample value available to us came from this population.

3. From this chance we decide whether the difference between sample and population is so unlikely as to be significant and hence whether our hypothesis is tenable.

Exercises

7.1 The marks over some years in the final school examination in mathematics have a mean of 60 and a standard deviation of 10. The 72 candidates from one school last year obtained a mean mark of 64. Can this school claim to be a superior school in mathematics?

7.2 A manufacturer of seat belts has found that over many years his product has had a breaking strength which is normally distributed with a mean of 600 kg and standard deviation of 25 kg. He has developed a new method of manufacture in order to increase the breaking strength and in a sample of 16 belts made by the new process he obtained the following breaking strengths:

608, 650, 624, 598, 645, 612, 595, 672, 643, 610, 599, 654, 642, 625, 596, 664.

Do these results indicate a significant increase in breaking strength?

7.3 A paint manufacturer has a standard test procedure for the durability of his products, in which he exposes specimens of the paint to controlled conditions and observes how long before the first signs of wear appear. He has on test at present two varieties of lilac paint known as Hydrangea and Heliotrope. Fifty samples of Hydrangea have taken an average time of 93.6 days to show first signs of wear, whereas fifty samples of Heliotrope have taken an average time of 90.9 days to show such signs. Past experience suggests that deterioration times have a standard deviation of nine days. Is this sufficient evidence to conclude that the variety known as Heliotrope deteriorates more rapidly than that known as Hydrangea?

7.4 The department of agriculture in a wheat producing country found in a very large survey that the diameter in millimetres of ears of wheat was nearly normally distributed with a mean of 40 mm and a standard deviation of 2.04 mm.

a. One farmer measured 64 ears selected at random from his crop and found the mean diameter to be 40.7 mm. Can he reasonably claim to produce a product larger than the national average?

b. Another farmer measured 49 ears selected at random from his crop and found the mean diameter to be 40.3 mm. Can the first farmer reasonably claim that his product is larger than that of the second?

8

The Binomial Distribution

A great many experimental situations consist of a succession of in-dependent trials, such that each trial must result in exactly one of two outcomes. For instance, in example *3* on page 65 each rat could either repeat or reverse and in example *1* on page 64 the acoustics engineer at each trial could be either right or wrong. Such a situation is called *binomial*. It is not necessary that the two outcomes should be equally likely; for instance each person in an examination must pass or fail, but presumably passing and failing are not equally probable.

To establish a terminology, it is convenient to describe one of the possible outcomes as 'success' and the other as 'failure', but it should be quite clearly understood that there need be no philosophical connotation to these words. Thus for instance, in a coin tossing situation, we may arbitrarily regard a tail as a success or a failure, so long as we adopt a consistent usage in any one problem.

We shall denote the probability of a success by π, the Greek equivalent of the Roman 'p'. (This usage has *nothing* whatever to do with the $\pi = 3.14159 \ldots$ often encountered in mathematics.) A theorem of mathematical statistics then states that if we observe n trials in a binomial situation, the number of successes obtained will have a theoretical population mean of $n\pi$ and standard deviation $\sqrt{n\pi(1 - \pi)}$. In other words, if we repeatedly observe sets of n trials and note the distribution of the number of successes, it would be characterised by that mean and standard deviation.

The examples discussed on pages 64 and 65 were all cases of a binomial distribution where $n = 100$ and $\pi = \frac{1}{2}$; the theoretical mean and standard deviation would thus respectively be $n\pi = 100 \times \frac{1}{2} = 50$ and $\sqrt{n\pi(1 - \pi)} = \sqrt{100 \times \frac{1}{2} \times \frac{1}{2}} = 5$. Thus, if we repeatedly observe the number of tails obtained in sets of 100 tosses of an unbiased coin, this number will vary around a typical value of 50 with a standard deviation of 5.

Example 8.1

Given that the probability of survival following a coronary occlusion is 0.8, how would the number of survivors out of groups of 400 vary?

Solution

This is a binomial situation because there are only two possible outcomes, death or survival. As we are studying the number of survivors, it is survival which, in our terminology, is described as a 'success'. As the number of trials is 400, the number of survivors will vary with a theoretical mean of 400×0.8 or 320 and standard deviation of $\sqrt{400 \times 0.8 \times 0.2}$ or 8.

Example 8.2

Given that the chance of an overhead projector pen being defective is 0.1, how would the number of defective pens in batches of 100 vary?

Solution

This again is a binomial situation because each pen is either defective or not defective. As we are studying the numbers defective, a 'success' in this case is selecting a defective pen. The number defective in batches of 100 will therefore vary with a theoretical mean of 100×0.1 or 10 and standard deviation of $\sqrt{100 \times 0.1 \times 0.9}$ or 3.

Example 8.3

Mendel's classic genetic experiment involved the crossing of two types of pea. The result of the crossing is the production of either yellow or green peas and, according to his theory the number of yellow and green peas should be in the ratio 3:1. If a number of lots of 1,200 peas obtained in this way were available, how would the number of yellow peas in each lot vary?

Solution

We are given that the resulting peas are either yellow or green, and hence we have a binomial situation. The probability that a pea is yellow is $\frac{3}{4}$. In lots of 1,200 the number of yellow peas would therefore vary with a theoretical mean of $1,200 \times \frac{3}{4}$ or 900 and standard deviation of $\sqrt{1,200 \times \frac{3}{4} \times \frac{1}{4}}$ or 15.

Normal approximation for the binomial

The coin tossing experiment described in Chapter 5 was a binomial situation and we pointed out that when the number of tosses of the coin in our experiment was large, say 100, the distribution of the number of tails obtained was not very different from the normal distribution discussed in Chapter 6. In fact the larger the number of tosses in this experiment the nearer does the distribution of the number of tails approach the normal distribution. This is another reason why the normal distribution tables are so useful in statistics.

For small values of n exact tables of the binomial distribution similar to *Table 5.1* are available in a number of books, and should be used. The tables are too elaborate to include in this book and we have therefore always used the normal tables as an approximation even for small values of n where it is not very exact. When π is near 0.5 the normal approximation is quite good even for quite small values of n, but when π is near 0 or 1 a larger value of n is required for an acceptable approximation.

Example 8.4

What would be the chance in *Example 8.3* that a lot chosen at random
(*i*) Had 927 or more yellow peas
(*ii*) Had from 880 to 927 yellow peas?

Solution
For such a large value of n we may assume that the number of yellow peas is normally distributed. The mean and standard deviation we have shown to be 900 and 15 respectively.

(*i*) The chance of obtaining 927 or more
    = the chance of obtaining 27 or more above the mean
    = the chance of a deviation of $\dfrac{27}{15}$ SDs or more above the mean
    = the chance of a standard score exceeding 1.80
    = 0.0359, from normal distribution tables.

(*ii*) The chance of obtaining 880 to 927 yellow peas
  = the chance of a deviation of from 20 below to 27 above the mean
  = the chance of a standard score from —1.33 to +1.80
  = 0.4082 + 0.4641
  = 0.8723, from normal distribution tables.

Testing hypotheses in binomial situations

We are frequently given information about a·sample taken from a binomial population and we wish to test some hypothesis about that population. A similar approach to that outlined in Chapter 7 may be adopted and normal distribution tables may be used if n is reasonably large.

Example 8.5

A gambler filed one edge of a coin in order to make it produce tails more often than heads. He checked the result of his work by tossing the coin 400 times. 221 tails resulted. Is he justified in concluding that the coin is now biased in favour of tails?

Solution
1. *Hypothesis* Assume that the coin is still unbiased. The number of tails should then be distributed with

$$\mu = \quad 400 \times \tfrac{1}{2} \quad = 200$$
$$\text{and } \sigma = \sqrt{400 \times \tfrac{1}{2} \times \tfrac{1}{2}} = \quad 10$$

2. *Deduction* The result obtained is 21 tails, or 2.1 standard deviations, above the mean. The use of normal distribution tables is justified because of the large value of n. From normal tables, the chance of obtaining a value 2.1 standard deviations or more above the mean is 0.0179. On the above hypothesis, 221 tails or more would occur by chance about 18 times in 1,000.

3. *Conclusion* There are two possible conclusions:
Either (*i*) our hypothesis is correct — the coin is still unbiased — and we have obtained one of those rare results which does occasionally happen by chance
or   (*ii*) our hypothesis is *not* correct and the coin is biased in favour of tails.

4. Comment The alternative (*ii*) seems the more reasonable conclusion. We could of course be wrong in this conclusion; in fact, we would expect to be wrong about 18 times in 1,000. This is a fairly small chance of being wrong and we may reasonably conclude that he has succeeded in biasing the coin in favour of tails.

Range of rejection

An approach to hypothesis testing which is frequently used is to determine a range within which acceptable sample values lie. This range is called the region of acceptance; values outside the range are in the region of rejection. The region of acceptance and the region of rejection vary with the level of significance used. Illustrations of this approach are given in *Examples* 8.6 (*ii*), 8.7 *b*. and 8.8.

Example 8.6

In an experiment to test Mendel's theory that crosses of two types of pea yield yellow and green peas in the ratio 3:1, 430 yellow peas and 126 green peas were obtained.

(*i*) If Mendel's theory is correct, what is the probability of obtaining by chance in a sample of this size a variation above or below the theoretical ratio of this amount or more?

(*ii*) In such an experiment, within what range should the number of yellow peas lie if, at the 5% level of significance doubts are not to be cast on the theory?

Solution

According to Mendel's theory the proportion of yellow peas in the population is $\frac{3}{4}$. In samples of 556 we would expect the number of yellow peas to vary almost normally with

$$\mu = 556 \times \tfrac{3}{4} = 417$$

$$\text{and } \sigma = \sqrt{556 \times \tfrac{3}{4} \times \tfrac{1}{4}} = 10.2$$

(*i*) A variation from the theoretical 3:1 ratio results if the number of yellow peas exceeds 417 or is less than 417. The variation in this experiment was 430 — 417 or 13. We are asked to find the chance of obtaining a variation from 417 of 13 or more. Now 13 equals $\dfrac{13}{10.2}$ standard deviations. We therefore require the chance of obtaining a standard score greater than 1.27 or less than —1.27. Using normal tables as an approximation, this probability equals $0.1020 + 0.1020 = 0.2040$

(*ii*) Reference to normal distribution tables shows that, if Mendel's theory is correct, in $2\frac{1}{2}\%$ of these experiments the number of yellow peas would be greater than 1.96 standard deviations above the mean and in $2\frac{1}{2}\%$ of the experiments the number of yellow peas would be less than 1.96 standard deviations below the mean. Now 1.96 standard deviations equals 1.96×10.2 or 20 peas. In only 5% of experiments therefore would we expect the number of yellow peas to exceed 437 or be less than 397 as a result of chance fluctuations. If the number in a particular experiment were to be outside this range then, at the 5% level of significance, doubts would be cast on the theory. In the language of statistics, at the 5% level of significance, the region of acceptance of the hypothesis is '397 to 437 peas' and the region of rejection 'more than 437 peas or less than 397 peas'.

Example 8.7

A doctor working in a foreign hospital knows that 80% of people treated in US hospitals for a particular disease recover. He has a treatment which he believes results in a higher recovery rate.

a. In the past two years, of 150 patients treated by him, 134 recovered. Does this result support his belief?

b. Determine for samples of this size the region of rejection of the null hypothesis

    (*i*) at the 5% level of significance

and (*ii*) at the 1% level of significance?

Solution

a. We follow the usual hypothesis testing approach of Chapter 7:

1. Hypothesis Assume that his treatment is really no better and that his apparently better results are just due to chance. The number of recoveries in samples of 150 patients would then have

$$\mu = 150 \times 0.8 = 120$$

$$\text{and } \sigma = \sqrt{150 \times 0.8 \times 0.2} = 4.9$$

2. Deduction The number of recoveries was 14 above the mean, i.e. $\dfrac{14}{4.9}$

or 2.86 standard deviations. Using the normal tables as an approximation since n is reasonably large, the chance of 134 or more recoveries is the chance of a standard score of 2.86 or more which equals 0.0021.

3. Conclusion There are two possible conclusions:

Either (*i*) our hypothesis is correct — his is *not* a better treatment — and we have obtained one of these rare results which does very occasionally occur by chance

or (*ii*) our hypothesis is *not* correct and his is really a better treatment.

4. Comment The alternative (*ii*) seems the more reasonable conclusion. We could of course be wrong in this conclusion; in fact, we would expect to be wrong about twice in 1,000 times. This is a very small chance of being wrong and we may reasonably conclude that the doctor has a treatment with a better recovery rate.

b. Our null hypothesis was that the recovery rate was 80%. With samples of 150 we have shown that on this hypothesis the mean number of recoveries is 120 and the standard deviation 4.9. In 5% of samples the number who recovered would *exceed* $120 + 1.645 \times 4.9$ or 128. In 1% of samples the number of recoveries would exceed $120 + 2.33 \times 4.9$ or 131.4. The region of rejection of the null hypothesis is thus

(*i*) at the 5% level, more than 128 recoveries
(*ii*) at the 1% level, more than 131 recoveries.

The result obtained in *a* namely 134 recoveries, lies in the region of rejection for either level.

One-tailed and two-tailed tests

In *Example 8.6(ii)* we were studying variations from Mendel's theory. An experiment in which the number of yellow peas produced was so great that the result could hardly have occurred by chance would lead us to suspect his theory. Equally, an experiment which produced so few yellow peas that the result could hardly have occurred by chance would also lead us to suspect his theory. As either too many or too few peas could lead us to suspect the theory, we used in our probability calculations *both* 'tails' of the normal distribution. Tests in which both 'tails' of a distribution are used are called two-tailed tests.

In *Example 8.7 a.* we were testing the doctor's belief that the recovery rate of his patients was *better*, not better or worse, than that of patients in US hospitals. In our probability calculations we therefore used only one tail of the normal distribution. Tests in which only one tail of a distribution is used are called one-tailed tests.

Example 8.8

A man believes that a particular six-faced die is unbiased with regard to the throwing of sixes. He decides to throw it 150 times as a check. At the 5% level of significance, within what range should the number of sixes lie if doubt is not to be cast on his belief?

Solution

Assume that the die is unbiased for sixes. The number of throws is again sufficient for the normal approximation to be used. The number of sixes may therefore be assumed to be normally distributed with

$$\mu = 150 \times \frac{1}{6} = 25$$

$$\text{and } \sigma = \sqrt{150 \times \frac{1}{6} \times \frac{5}{6}} = 4.55$$

In 5% of samples the number of sixes would lie outside the range $25 \pm 1.96 \times 4.55$, that is, outside the range 16.1 to 33.9.

If the number of sixes thrown lies between 16 and 34 we have no reason, at the 5% level of significance, to doubt the hypothesis that the die was unbiased for sixes. In statistical language, the region of acceptance at the 5% level is between 16 and 34 sixes.

As the die could be biased either in favour of too many or of too few sixes a two-tailed approach to the problem was necessary. If in problems of this type we are given the result of a particular experiment and we are asked to test a given hypothesis, the decision as to whether to use a two-tailed or a one-tailed test depends on the particular question we wish to answer.

Possible questions are: With regard to the throwing of sixes is the die (*i*) unbiased (*ii*) biased in favour of sixes (*iii*) biased against sixes? A two-tailed test would generally be used for (*i*), and a one-tailed test for (*ii*) and (*iii*).

Continuity correction

Fig. 5.4 (page 61) presents probability distributions in a number of binomial situations. It is clear that the distributions proceed in 'jumps'. This is a result of the fact that the number of tails must be an integer. To obtain 45.3 tails, for example, is not possible. The normal probability distribution illustrated in *Fig. 6.1* (page 68) is what is called a continuous distribution. It moves smoothly from point to point and fractional values *are* possible. If *Table 5.1* (page 63) were not available and we wished to find the probability of obtaining 46 heads we could use, as explained earlier, the normal distribution tables as they provide a good approximation. In this case with 100 throws and $\pi = \frac{1}{2}$ we have

$$\mu = 100 \times \tfrac{1}{2} = 50$$

and $\sigma = \sqrt{100 \times \tfrac{1}{2} \times \tfrac{1}{2}} = 5$

We wish to obtain the area, in the last histogram of *Fig. 5.4*, of the bar standing above the value 46. The corresponding normal curve would pass through the mid-points of the top of each bar in the histogram. As all values, not merely integers, are possible with a normal curve the scale on the x-axis is a continuous scale and the base of the particular bar of the histogram we are considering corresponds in the case of a normal curve, to the values from $45\frac{1}{2}$ to $46\frac{1}{2}$. The probability of 46 heads is thus approximately the probability of a value from $45\frac{1}{2}$ to $46\frac{1}{2}$ under the normal distribution with mean 50 and SD 5. That is, it is the probability of a standard score between $-\dfrac{4\frac{1}{2}}{5}$ and $-\dfrac{3\frac{1}{2}}{5}$ or between

-0.9 and -0.7. From normal tables this probability equals

$$0.3159 - 0.2580 = 0.0579$$

The exact value from *Table 5.1* is 0.05796.

In the same way the probability of obtaining from 46 to 48 tails (inclusive) equals the probability of obtaining values between $45\frac{1}{2}$ and $48\frac{1}{2}$ from the same normal distribution. This equals the probability of a standard score between -0.9 and -0.3 which, from normal tables, equals

$$0.3159 - 0.1179 = 0.1980$$

The exact value from *Table 5.1* is 0.19808.

Finally, the probability of obtaining 58 or more tails equals the probability of obtaining a value of more than $57\frac{1}{2}$ from the same normal distribution. This equals the probability of a standard score greater than $\dfrac{7\frac{1}{2}}{5}$ or 1.5 which equals 0.0668.

This adjustment of $\frac{1}{2}$ which should be made before entering the normal tables is called the continuity correction. The use of the continuity correction frequently makes little difference to the result obtained. We did not in fact use it in the examples dealt with in this chapter. It is however simple to apply; it is theoretically correct; it should therefore be used as a matter of routine.

Example 8.9

A binomial population has a probability of success of 0.4. A sample of size 50 is selected at random. Find the chance that the number of successes
  (*i*) is 16 or less
 (*ii*) is 26 or more
(*iii*) lies between 16 and 26 (excluding both numbers).

Solution

$$\mu = \quad 50 \times 0.4 \quad = 20$$

$$\sigma = \sqrt{50 \times 0.4 \times 0.6} = 3.46$$

(*i*) The probability required is obtained by using normal distribution tables and calculating the probability of a value of $16\frac{1}{2}$ or less. This is the probability of a standard score less than $-\dfrac{3\frac{1}{2}}{3.46}$ or -1.01, namely $0.5 - 0.3438 = 0.1562$.

(*ii*) The probability required is obtained by using normal distribution tables and calculating the probability of obtaining a value of $25\frac{1}{2}$ or more. This is the probability of a standard score greater than $+\dfrac{5\frac{1}{2}}{3.46}$ or $+1.59$, namely $0.5 - 0.4441 = 0.0559$.

(*iii*) The probability required is obtained by using normal distribution tables and calculating the probability of obtaining a value between $16\frac{1}{2}$ and $25\frac{1}{2}$. This is the probability of a standard score between -1.01 and $+1.59$, namely $0.3438 + 0.4441 = 0.7879$.

Note: The sum of these three answers, namely $0.1562 + 0.0559 + 0.7879$, equals 1 since one of the three possibilities must occur.

Exercises

8.1 If every student in a class were to perform the experiment of tossing an unbiased coin 1,000 times and recording the number of heads, what would be the mean and standard deviation of the collection of scores obtained?

8.2 If every person in this class were to perform the experiment of throwing an unbiased six faced die 1,200 times and recording the number of fours, what would be the mean and standard deviation of the collection of scores obtained?

8.3 Past experience shows that in first year english 50% of students pass. In a new class of 200 what is the probability that at least 55% will pass?

8.4 If 30% of students in NSW are medically unfit (i.e. have poor eyesight, flat feet etc.) what is the probability that at least 35% of the members of a particular school of 300 are medically unfit?

8.5 A manufacturer of washing machines has found that 10% of his machines break down within the first year of service.
a. What is the probability that a laundrette which has just bought 100 of his machines will find at least 18 of them break down within a year?
b. What is the probability that at least 5 and not more than 15 will break down?

8.6 The standard cure for a certain disease is successful in 20% of cases. A new treatment is tried on 200 patients and is successful in 60 cases. Can we conclude without doubt that it is a more successful treatment?

8.7 A professor decides to compare the pass rate in psychology this year with the rates of previous years. He finds that over a long period of time the percentage passing has been consistently 75%. Of his class of 400 this year, 290 passed. Could this poorer performance be reasonably attributed to chance?

9

Revision of Chapters 6 to 8

This chapter is a revision of Chapters 6, 7 and 8. We have there mentioned that the following are examples of populations which are approximately normal and for which the standard normal tables can be used. (We shall use the notation $N(a, b)$ to indicate the normal distribution with mean a and standard deviation b.)

1. Heights of individuals, weights of individuals, measurement errors, yields of crops, intelligence quotients, heights or distances jumped

2. The means of fairly large samples from *ANY* population; their distribution can be regarded as $N\left(\mu, \dfrac{\sigma}{\sqrt{n}}\right)$

3. The number of 'successes' in large binomial populations, for example dice throwing, coin tossing, numbers voting Liberal; here the distribution is $N\left(n\pi, \sqrt{n\pi(1-\pi)}\right)$

4. The difference between the means of two independent samples of size n_1 and size n_2 from normal populations with the same mean and with standard deviations σ_1 and σ_2 respectively; the distribution is

$$N\left(0, \sqrt{\frac{\sigma_1{}^2}{n_1} + \frac{\sigma_2{}^2}{n_2}}\right).$$

Because of the Central Limit Theorem, the same formula may be used if the populations are not normal, provided the samples are large.

Problems are usually of two types:
A. Where we are given full information regarding the population and

are asked to find the probability of obtaining by chance a given sample from it; and

B. Where we are given the results of a sample and we want to test some assumptions which we make about the population from which this sample was taken.

Example 9.1

Category 1A. Given that the heights of men are normally distributed with mean 175 cm and SD 8 cm , what is the chance that a person selected at random from the population has a height

a. Greater than 181 cm ?

b. Between 171 cm and 183 cm.

Solution

a. Persons with heights greater than 181 cm have heights 6 cm or $\frac{3}{4}$ of a standard deviation above the mean. From normal distribution tables the probability of obtaining by chance a single individual with such a height is .5 — .2734 = .2266 or 22.66% or 22.66 in 100.

b. In this case it is the probability of obtaining a single individual with height between $\frac{1}{2}$ SD below and 1 SD above the mean. This probability is .1915 + .3413 = .5328.

Example 9.2

Category 1A. Individual marks in the state in english are distributed normally with mean 58 and SD 10. How many out of a class of 200 would be expected to obtain 75 or more?

Solution

Seventy-five marks is 17 above the mean, i.e. $\dfrac{17}{10}$ or 1.7 SDs above the mean. The probability of getting 1.7 SDs or more above the mean is 0.0446. The number expected to obtain 75 or more is therefore $200 \times .0446 \doteqdot 9$.

Example 9.3

Category 2B. The weights of 16 year old boys are distributed normally with mean 70 kg and SD 8 kg. At a certain boys' school, boarding 100 16 year olds, the mean weight was 67 kg. Comment.

Solution

If the weights of individual 16 year olds are normally distributed with $\mu = 70$ and $\sigma = 8$ then the mean of groups of 100 will be normally distributed with mean 70 and SD $\dfrac{8}{\sqrt{100}}$ or 0.8. The 'population' here consists of a large number of 'means of groups of 100'. This population has a mean of 70 kg and SD 0.8 kg. We have one value which, if it comes from this population, differs by 3 kg , i.e. $\dfrac{3}{0.8}$ or 3.75 standard deviations from the mean. The chance of obtaining a value of 67 kg or less is thus less than 1 in 1,000. It is most unlikely therefore that this is a random sample from the population. It is more likely that the boys at this school are fed differently.

Example 9.4
Category 2B. Individual marks in the state in english are distributed normally with mean 58 and SD 10. The average mark of 196 boys from one school was 62. Can this school claim to be superior in english?
Solution
Since we are given that individual marks are N (58, 10), the mean of groups of 196 will be normally distributed with mean 58 and SD $\dfrac{10}{\sqrt{196}}$ or $\dfrac{10}{14}$. The population here is 'means of groups of 196' and we have one sample which differs from the population mean by 4 which is $4 \div \dfrac{10}{14}$ or 5.6 standard deviations. We cannot read from our tables the probability of a deviation of this amount or greater but it is clearly much less than 1 in 1,000. Hence we conclude that it is almost certainly not a random sample from the population. The school can reasonably claim to be superior in english.

Example 9.5
Category 1B. An anthropologist on safari in New Guinea meets a native of some new tribe. He is 165 cm in height. Is he justified in saying that the new tribe is of smaller stature than Australians whose mean height is 175 cm with SD 8 cm ?
Solution
We assume that there is no difference between the heights of this tribe and those of Australians, that is that they all belong to the same

population of heights. The height of this native is then a random sample from the population and his height differs by 10 cm i.e. $\frac{5}{4}$ SDs from the population mean. The chance of obtaining a value of 165 cm or less is thus .1056 or about 1 in 10. This is not an unlikely occurrence and we cannot from this case deduce that the new tribe is of smaller stature than Australians.

Example 9.6
Category 2B. He later meets 24 other natives of the same tribe and the mean height of the 25 is 169 cm. What can he now conclude?
Solution
We again assume that there is no difference between the height of these natives and of Australians, that is they belong to the same population with mean 175 cm and SD 8 cm. If we study groups of 25, then the means of the groups would be normally distributed with mean 175 cm and SD $\frac{8}{\sqrt{25}}$ or $\frac{8}{5}$. Our particular sample has a value of 169 cm which on our assumption is from a population with mean 175 cm and SD $\frac{8}{5}$ cm. Our sample is 6 cm or $3\frac{3}{4}$ standard deviations below the mean.

The probability of getting by chance a sample with this mean height or less is less than .001. This is most unlikely and we conclude that our assumption is *not* valid. The anthropologist can safely conclude that the tribe is shorter than Australians.

Example 9.7
Category 3A: What is the chance of obtaining 110 or more sixes in 600 throws of an unbiased six-faced die?
Solution
With binomial samples of this size the number of sixes would be normally distributed with mean $600 \times \frac{1}{6}$ and SD $\sqrt{600 \times \frac{1}{6} \times \frac{5}{6}}$ i.e. $\mu = 100$, $\sigma = 9.13$. Applying the continuity correction we obtain a deviation of $9\frac{1}{2}$. The probability we require is therefore the probability of obtaining a normal deviation exceeding $\frac{9.5}{9.13}$ or 1.04 standard deviations. This probability equals 0.1492 or about 15 in 100.

Example 9.8
Category 3B. In a game of dice A lost rather badly. He therefore carried out an experiment with the particular die used and found that, in 600 throws, a 1 was thrown 150 times. What conclusion should he draw?

Solution
Assume the die was unbiased. Then, if the experiment A carried out had been carried out many times, the number of ones thrown would be normally distributed with mean 100 and SD 9.13 (see previous problem). The chance of 150 ones or more is the same as that of a normal deviation exceeding $\dfrac{49\frac{1}{2}}{9.13}$ or 5.4 standard deviations. We cannot read this probability from our tables but it is clearly much less than 1 in 1,000. This chance is so small that A must conclude his assumption was wrong and that the die is strongly biased in favour of one.

Example 9.9
Category 1A. Experience of scholarship examinations over many years showed that the marks of candidates are approximately normally distributed with a mean of 60 and standard deviation 10. As a matter of policy, we must specify in advance what the pass mark will be, but because of financial stringencies, we can only offer approximately 10 scholarships. If we know that 200 students will be sitting for the examination, what qualifying mark should we set?

Solution
The proportion of the population to be above the selected mark is 10/200 (or 0.05 or 5%). Therefore we want to select a mark so that the probability of a student getting a mark above this is 0.05. From the tables, we see that a mark which is 1.645 standard deviations above the mean has this property, and we will therefore choose $60 + 10 \times 1.645$, i.e. 76 approximately, as our qualifying mark.

Example 9.10
Category 4B. A research worker carrying out experiments on the growth of rats over a specific short period from birth has found the standard deviation of the weight of individual rats to be 2 grams. He is carrying out experiments to test the effect of two diets on growth. He selects 100 rats, divides them at random into two groups of 50, and feeds one group with diet A, the other with diet B. The mean weight of those on

diet A was 14.4 grams and of those on diet B, 13.4 grams. Can he conclude that diet A is more effective in stimulating growth?

Solution

Assume that diet has no effect and that these are samples drawn from the same population, the difference being solely due to chance. The difference of the means of the samples (on this assumption) is normally distributed with mean zero and $\text{SD} = \sqrt{\dfrac{4}{50} + \dfrac{4}{50}} = 0.4$ grams. (The normal distribution may be assumed as such weights are near normal and in any case the sample is large). In this experiment the difference found between the means is 1.0 grams. Since this difference on our assumption is distributed as $N\ (0,\ 0.4)$ we have obtained a standard score of $\dfrac{1.0}{0.4}$ or 2.5. The probability of obtaining this standard score or more by chance is 0.006 or only 6 in 1,000. This is such an unlikely event that he must doubt the hypothesis and conclude that diet has a real effect.

Note: The mean weight of the 100 rats is 13.9 grams. The mean of neither sample differs significantly from this, yet the difference between the means of the samples is quite significant.

10

The *t* Test

Testing means of samples when the population SD is unknown

It is now appropriate to point out that when earlier we were given a sample from a population and were testing to see whether a sample with this mean could have come from a given population we always assumed that the population standard deviation was known. Very often this is not the case. There are, however, two ways of overcoming the difficulty.

(1) Large samples — any population

With large samples (say over 50) we may use two reasonable approximations:

a. Because of the Central Limit Theorem we may consider the mean of the sample to be approximately normally distributed, whatever the population distribution;

b. We may use the variation within the sample as an indication of the variation within the population. It is sufficiently accurate for testing purposes, in samples of this size, to use the sample standard deviation s as an approximation for the unknown population standard deviation σ. Given $\bar{x}$ and s for such a sample, we may therefore use normal tables and the methods already developed for the case where the population standard deviation is known, to test whether this sample could reasonably have come from a population with some given mean.

Example 10.1

The scores obtained in a manual dexterity test by pupils entering secondary schools have, over a long period, averaged 100. A new group of 50 had a mean value of 103.5 and a standard deviation of 20. Have they significantly greater dexterity?

Solution

1. Hypothesis Assume that their dexterity is the same as that of former students.

2. Deduction We are therefore assuming that this group is a random sample from a population whose mean is 100. Since the sample is large we use the sample SD of 20 as an approximation for the population SD. The mean manual dexterity of a sample of 50 should therefore be normally distributed with mean 100 and SD $\dfrac{20}{\sqrt{50}}$ or 2.83. The new group has a mean of 103.5 which is equivalent to a standard score of $\dfrac{3.5}{2.83}$ or 1.24. The chance of a standard score of 1.24 or greater is 0.1075.

3. Conclusion This is not a very unlikely event. The hypothesis is thus not disproved and we cannot conclude that the new group has higher manual dexterity. The variation may be explained by chance.

(2) *Small samples — normal populations*

If the sample is small we cannot reasonably use the variation within the sample as a precise indication of the population standard deviation. We cannot therefore calculate a standard score and apply the normal distribution tables. We can however perform an analogous operation and calculate a quantity known as Student's *t* which is based on the *sample* standard deviation and does not require any information about the *population* standard deviation. Tables of the *t* distribution have been prepared for samples *drawn from normal populations*. The *t* distribution tables are exact, however small the sample, provided the population is normally distributed. In practice they can be used for samples, no matter how small, provided the population is normal or close to normal.

We define *t* as

$$t = \frac{\bar{x} - \mu}{s/\sqrt{n-1}} \quad \text{or} \quad \frac{\bar{x} - \mu}{s}\sqrt{n-1}$$

i.e., it is the difference between the observed sample mean and the hypothetical population mean, divided by (the standard deviation of the sample $\div \sqrt{n-1}$).

A moment's thought will show that the standard score we calculate in the situation where σ is known is actually the quantity

$$\frac{\bar{x} - \mu}{\sigma/\sqrt{n}}$$

so that the t statistic is of generally similar form although different in detail.

t distribution tables (see page 257) usually specify the values of t which are exceeded in (say) 1 random sample in 10, 1 random sample in 20, 1 in 100 etc. The table on page 257 is set out in a form convenient for a two-tailed test but the necessary information for a one-tailed test can easily be obtained from it. The value of t which is exceeded with a given probability varies with the size of the sample. The values of t are usually given in the tables for each value of $n - 1$, rather than the number n in the sample. This quantity $n - 1$ which appears in the formula for t, is referred to as the number of degrees of freedom.

Reference to the tables will show that, if we are using a two-tailed test for a sample of size 11, i.e. with 10 degrees of freedom, values of t numerically greater than 2.23 (i.e. greater than $+2.23$ or less than -2.23) will occur in 5 random samples in 100. In the case of a one-tailed test with 10 degrees of freedom values of t greater than $+1.81$ will occur in 5% of samples. In this latter case we have used the 0.10 column as the table is set out for a two-tailed test; for a one-tailed test the corresponding probability is one half of this, namely 0.05.

The t distribution is like the normal in shape and in fact approaches the normal as the number of degrees of freedom becomes large.

Note:

1. There is considerable similarity between methods (1) and (2).

2. Method *(1)* applies only to large samples but does not require normal populations, because of the Central Limit Theorem.

3. Method *(2)* is accurate for samples however small, but in theory requires normal populations and in practice populations close to normal.

Example 10.2

What conclusions could we have drawn from the data in *Example 10.1* if the number of pupils in the new group had been 15?

Solution

(*i*) Assume again that the new group have the same dexterity as former groups. We are not told in the question that the scores followed a normal distribution but experimental results have generally shown such test scores to have a near normal distribution. It is reasonable therefore to assume that this is a sample from a normal population and to apply a t test.

$$(ii) \qquad t = \frac{103.5 - 100}{20} \sqrt{14} = 0.65$$

(*iii*) With 14 degrees of freedom, tables of the t distribution show that, with a one-tailed test at the 5% significance level, values of t up to 1.76 may be expected.

(*iv*) This sample is thus well within the range of variation due to chance and hence we cannot deduce that this group has higher manual dexterity.

Example 10.3

Assume we know from past experience that the gain in weight of infants over a one-month period is normally distributed about a mean of 250 grams. A doctor tries a new diet on a sample of 17 infants and finds the mean gain of the sample to be 312 grams and its value of s 100 grams. Is the new diet really more effective?

Solution

  (*i*) Assume that it is not.

  (*ii*) The population is normal and for the sample

$$t = \frac{312 - 250}{100} \sqrt{16} = 2.48$$

(*iii*) With 16 degrees of freedom, tables of the t distribution show that, with a one-tailed test at the 5% significance level, values of t up to 1.75 may be expected.

 (*iv*) The value produced by this sample is therefore unlikely to have arisen from chance alone. We thus reject the hypothesis and presume the diet is really effective.

Testing the difference between means of samples: population variance unknown

If we select two independent random samples from a normal population whose mean and standard deviation are unknown then it can be shown that the difference between the means of the samples will be normally distributed with mean zero and some unknown standard deviation. The *t* distribution will therefore apply to these differences and may therefore be used to test whether two samples are likely to have been drawn from the same normal population.

In this case the values to be used in the formula for *t* are as follows (the subscripts 1 and 2 refer to the two samples):

For $\bar{x} - \mu$ use $\bar{x}_1 - \bar{x}_2$

For s use $\sqrt{\dfrac{n_1 s_1{}^2 + n_2 s_2{}^2}{n_1 + n_2 - 2}}$

For $\sqrt{n-1}$ use $\sqrt{\dfrac{n_1 n_2}{n_1 + n_2}}$

The number of degrees of freedom equals $n_1 + n_2 - 2$.

Example 10.4

We have tested two samples of light bulb for length of burning time.

10 of Type A gave $\bar{x}_1 = 1{,}000$ hours and $s_1 = 80$ hours
12 of Type B gave $\bar{x}_2 = 950$ hours and $s_2 = 75$ hours

Is Type A superior to Type B?

Solution

(*i*) Assume these are samples from the same normal population.

(*ii*) $t = \dfrac{1000 - 950}{\sqrt{\dfrac{10 \times 80^2 + 12 \times 75^2}{10 + 12 - 2}}} \sqrt{\dfrac{10 \times 12}{10 + 12}} = 1.44$

(*iii*) Number of degrees of freedom $= 10 + 12 - 2 = 20$

(*iv*) With 20 degrees of freedom, tables of the *t* distribution show that, with a one-tailed test at the 5% significance level, values of *t* up to 1.72 may be expected.

(*v*) We cannot therefore conclude on this evidence that A is superior to B.

Example 10.5

Drug A was administered as an experiment to 10 healthy subjects who had been kept for a week in bed in hospital as a prelude to the experiment, and the hours of sleep (x_1) which followed were recorded; drug B was administered to another 10 patients in similar circumstances and the hours of sleep (x_2) in their case also recorded. The results obtained were:

$$\bar{x}_1 = +10.8 \text{ hours} \qquad s_1 = 1.2 \text{ hours}$$
$$\bar{x}_2 = +12.4 \text{ hours} \qquad s_2 = 1.0 \text{ hours}$$

Is there any difference in the effectiveness of the two drugs as measured by the resulting hours of sleep?

Solution

(*i*)   We assume that the number of hours of sleep produced is normally distributed. In practice we would need to check results obtained in earlier experiments to ensure that this is not an unreasonable assumption. We also assume the null hypothesis that there is no difference in the effectiveness of the two drugs in producing sleep, that is we assume that these are two random samples from the same normal population.

$$(ii) \qquad t = \frac{12.4 - 10.8}{\sqrt{\dfrac{10 \times 1.2^2 + 10 \times 1.0^2}{10 + 10 - 2}}} \sqrt{\frac{10 \times 10}{10 + 10}} = 3.20$$

(*iii*) Number of degrees of freedom $= 10 + 10 - 2 = 18$.

(*iv*) With 18 degrees of freedom, tables of the t distribution show that, with a two-tailed test at the 5% significance level, values of t up to 2.10 may be expected. We chose to use a two-tailed test because the question asked whether there is *any difference* between the sleep-producing effects of the two drugs, not whether drug B is *more effective* than drug A.

(*v*) Since a value of t of 3.07 was obtained from this experiment we may conclude, provided our assumption of a normal distribution is reasonable, that there is a real difference in effectiveness between the two drugs in producing sleep.

Paired tests

In this last example (*Example 10.5*) the two drugs were administered to 20 different patients. We therefore compared the effect produced by one drug on 10 patients with the effect produced by the other drug on 10 *different* patients. It is well known that people vary greatly in their reaction to drugs, and some of the difference found in this experiment could have been due to differences in patient's reaction rather than the direct result of differences in the drugs. These personal variations in response to drugs could have been eliminated from the experiment and more reliable results obtained if we had administered drug A, and on another occasion drug B, to the *same* patient and studied the *difference* in the hours of sleep produced. If there is no difference in the effects of the drugs then the theoretical mean of the differences should be zero. The *t* test may be used to test whether the average difference obtained in a particular experiment differs significantly from zero, *provided* the differences can be assumed to be normally distributed. Such a test is called a paired *t* test.

Example 10.6

The hours of sleep experienced by 10 healthy subjects in the circumstances outlined in *Example 10.5* following the use of two new drugs were as follows:

Subject	1	2	3	4	5	6	7	8	9	10
Drug C	+10.5	+8.8	+10.0	+10.1	+10.3	+12.1	+11.4	+11.2	+9.2	+11.7
Drug D	+11.5	+11.0	+11.1	+11.2	+10.1	+12.9	+13.0	+11.8	+13.6	+12.9

Care was taken to ensure that the background circumstances and routine for each subject were the same preceding the taking of drug C as they were preceding the taking of drug D.

Is there any difference in the effectiveness of the two drugs as measured by the resulting hours of sleep?

Solution

(*i*)   We make the null hypothesis that there is no real difference in the sleep-producing effect of the two drugs. As in *Example 10.5*, we also assume that the number of hours of sleep produced is normally distributed. Strictly, the differences only need to be normally distributed.

(*ii*) The actual differences are:

$$1.0, \quad 2.2, \quad 1.1, \quad 1.1, \quad -0.2, \quad 0.8, \quad 1.6, \quad 0.6, \quad 4.4, \quad 1.2$$

On our hypothesis this is a random sample from a normal population with a mean of zero. From these figures

$$\bar{x} = 1.38 \qquad s = 1.167$$

$$\text{giving } t = \frac{1.38 - 0.0}{1.167} \sqrt{9} = 3.55$$

(*iii*) Number of degrees of freedom $= 10 - 1 = 9$.

(*iv*) With 9 degrees of freedom, tables of the *t* distribution show that, with a two-tailed test at the 5% significance level, values of *t* up to 2.26 may be expected.

(*v*) We conclude that there is a difference in the sleep-producing effects of the two drugs.

Exercises

10.1 The mean mark in a certain examination was 62. A certain tutor who had a group of 17 students found that their results were 64, 67, 69, 70, 68, 65, 60, 58, 64, 60, 60, 65, 67, 65, 64, 63 and 62. Does this group appear to be significantly better than the students as a whole?

10.2 A plumber keeps a record of all the useless small pieces of copper pipe which he acquires as leftovers from longer pieces, and has noted over the years that these offcuts have a mean length of 18 cm. He has been keeping a watchful eye on one of his apprentices, however, and notices that the 37 pieces produced by the apprentice in one week had a mean length of 22 cm , with standard deviation 6 cm. Do you feel that he is justified in reprimanding the apprentice for inefficient use of the pipes?

10.3 Fifty school children were given a series of quick addition tests under normal conditions and the mean mark obtained was 93.6 with a standard deviation of eight.
Two days later they were given a similar test with a loud background of music. Under these conditions the mean mark was 90.9 and standard deviation ten.
Do quiet conditions make for better performance?

10.4 Twelve houses were painted as an experiment. Part of each house was given one undercoat and part given two. For each of the twelve houses the number of months which elapsed after painting before repainting was considered necessary is set out below:

2 undercoats	52	42	51	72	60	58	57	50	70	36	46	35
1 undercoat	42	39	48	50	42	60	29	38	62	46	42	41

Does the second undercoat produce a more lasting result?

11

The χ^2 Test

If an unbiased six-faced **die** is thrown many times the actual number of sixes thrown should be quite close to one-sixth of the total number of throws. This 'one-sixth of the total number of throws' is described in statistical language as the expected number of sixes. Even with an unbiased **die** the number of sixes obtained in an experiment will of course usually depart from this theoretical or expected number and the departure could be relatively large if the total number of throws is small.

In Chapter 8 we tested whether a six-faced **die** was biased by comparing the actual number of throws recording six with the number expected. A **die** could however be unbiased for throws of six but favour, say two, at the expense of four. To test whether the **die** was completely unbiased we would need to compare the actual frequencies of occurrence of each number, one to six, with the expected. The results of sixty throws might be as follows:

	1	2	3	4	5	6	Total
Observed	14	6	5	13	5	17	60
Expected	10	10	10	10	10	10	60

One way of measuring the difference between the actual result of the experiment and the expected would be to calculate the quantity called chi-squared (written χ^2) which is defined as

$$\chi^2 = \Sigma \, \frac{(\text{actual frequency} - \text{expected frequency})^2}{\text{expected frequency}}$$

In this example

$$\chi^2 = \frac{4^2}{10} + \frac{(-4)^2}{10} + \frac{(-5)^2}{10} + \frac{3^2}{10} + \frac{(-5)^2}{10} + \frac{7^2}{10} = 14.0$$

If the observed frequencies and expected frequencies were identical, χ^2 would clearly be zero. Obviously it cannot be negative. Also, the larger the discrepancy the larger the value of χ^2.

Assuming the differences between actual and expected are due to chance alone, the values of χ^2 which are likely to occur have been calculated and are tabulated in the *Appendix* (page 258). The larger the value of χ^2 the less likely is it to have occurred by chance. The tables give the values of χ^2 which would be exceeded by chance with given probability i.e. in a given proportion of cases. Also, the more categories we have the larger we would expect χ^2 to be. In fact, for a given probability level, the values of χ^2 depend on what is called the number of degrees of freedom. This is the minimum number of categories for which data is required to enable the full results of the experiment to be presented. For example, if we know the number of throws showing 1 to 5 we could fill in the number showing 6 by subtraction from the total of sixty. There are thus a minimum of 5 categories for which data must be given i.e. there are 5 degrees of freedom.

Tables of χ^2 show that at the 5% level of significance for five degrees of freedom $\chi^2 = 11.07$. This means that, on the hypothesis on which our expected values were based, in 19 random samples out of 20 the value of χ^2 will be less than 11.07. A value as large as 14.0, as obtained in our experiment, would therefore only occur by chance less than one time in twenty. We may therefore well suspect our hypothesis and thus suspect that the **die** is biased.

The number of degrees of freedom relating to a particular value of χ^2 is often indicated by recording that number as a subscript to χ^2. Thus, in the above case we found a value of χ^2_5 of 14.0.

Let us consider two other examples.

Example 11.1

In Mendel's classic pea breeding experiment referred to in Chapter 8, the results of crossing were as follows:

Round and yellow	315
Wrinkled and yellow	101
Round and green	108
Wrinkled and green	32

Mendelian theory predicts that they should be in the ratio 9 : 3 : 3 : 1. Are these results compatible with the theory?

Solution

The total number of observations = 556.

The expected numbers in the various categories are:

9/16 of 556, 3/16 of 556, 3/16 of 556, and 1/16 of 556

or   312.75, 104.25, 104.25, 34.75

Thus $\chi^2 = \dfrac{(2.25)^2}{312.75} + \dfrac{(3.25)^2}{104.25} + \dfrac{(3.75)^2}{104.25} + \dfrac{(2.75)^2}{34.75} = 0.47$

If three of the observed values are known, the fourth is fixed. Hence there are three degrees of freedom. χ^2 at the 5% level equals 7.8 for three degrees of freedom and hence this result does not contradict the theory.

Contingency tables

When carrying out a survey or an experiment the research worker is frequently interested in two characteristics or attributes of the individuals who are the subject of the research. He may be studying, for example, the education level of women with completed families and the number of children born to them, or in the kind of soap used by an individual and his income. The results of such surveys are frequently presented in the form of two-way tables called contingency tables. The following is such a table which sets out the hypothetical results of a survey into the education level of 500 persons and the extent of their use of public libraries:

Education Level	Use of public libraries				
	No use	Some use	Irregularly but quite often	Frequent and regular use	Totals
Tertiary	30	35	120	15	200
Secondary	25	45	50	30	150
Primary	25	40	30	55	150
Totals	80	120	200	100	500

This data was presumably obtained to determine whether there is any connection between education level and the use of public libraries. The χ^2 test may be used to determine whether there is any such connection or whether the use of public libraries by individuals is independent of their education level.

The analysis of the data proceeds as follows:

(*i*) Assume that there is no connection between education level and the use of public libraries.

(*ii*) Of the 500 persons, 80 made no use of public libraries.

On our assumption we would expect this same proportion $\left(\dfrac{80}{500}\right)$ of those who received tertiary education (200) to make no use of public libraries. Hence the expected number (in the sense mentioned at the beginning of this chapter) corresponding to the actual number of 30 in the top left-hand corner of the table is $\dfrac{80}{500} \times 200 = 32$. Similarly, the expected numbers corresponding to 35 and 120 in the same line are 48 and 80. The expected number corresponding to 15 in that line can be determined by subtraction since the total number who received tertiary education was 200. This number equals $200 - 32 - 48 - 80 = 40$.

In this way the following table of *expected* frequencies was calculated:

32	48	80	40	200
24	36	60	30	150
24	36	60	30	150
80	120	200	100	500

We now have both the actual frequencies and the expected frequencies and can therefore substitute in the formula for χ^2. We thus obtain

$$\chi^2 = \frac{(30-32)^2}{32} + \frac{(35-48)^2}{48} + \ldots + \frac{(30-60)^2}{60} + \frac{(55-30)^2}{30}$$
$$= 79.5$$

(*iii*) Although there are 12 items to be added to give χ^2, 12 is not the number of degrees of freedom. Once we have worked out the six items within the broken lines the other items used in determining χ^2 can be obtained by subtraction. Hence there are six independent values and hence six degrees of freedom.

(*iv*) At the 5% level with 6 degrees of freedom the χ^2 table shows that values of χ^2 up to 12.6 may be expected by chance.

(*v*) The value of χ^2 obtained from this hypothetical survey is well above this and hence we must doubt our hypothesis that there is no connection between education level and the use of public libraries.

Note: The above example illustrates the following general results which apply to all contingency tables:

1. The number of degrees of freedom $= (m - 1)(n - 1)$ where m is the number of rows in the table and n the number of columns.

2. For each row and for each column the sum of the expected numbers must equal the sum of the observed numbers.

Limitation of the χ^2 test

The χ^2 test is liable to give misleading results if any expected frequency is less than five. If the frequency in any cell is less than five it is usual to combine this cell with another before applying the test.

Combining several experiments

If an experiment has been carried out to test a certain hypothesis and a value of χ^2 with a given number of degrees of freedom has been calculated, it is possible that a significant result may not be obtained. If the experiment is repeated independently several times it can be shown that the values of χ^2 may be added and the total χ^2, from all the experiments, may be used combined with the total of the degrees of freedom, as a test of the hypothesis.

Example 11.2

G. N. Pollard[1] investigated the relationship between the ages of Australian mothers and the sex ratio of their children. On the assumption that these are independent of one another he obtained the following values of χ^2, each with one degree of freedom:

Period	χ^2_1
1914–17	5.98
1918–22	0.54
1923–27	2.37
1928–33	2.52
1934–38	8.29
1939–43	1.35
1944–48	5.83
1949–53	3.15
1954–58	4.03
1959–63	1.53

[1] G. N. Pollard, 'Factors influencing the sex ratio at birth in Australia 1902-65', *Journal of Biosocial Science*, vol. 1, 1969, p.125

Do these values indicate significant variation in sex ratio with age of mother?

Solution

In only one period, namely 1934–38, did the value of x_1^2 exceed the 1% significance level. That is, only one of these values is outside the range of values of x_1^2 which occur by chance in 99 samples out of 100. We cannot therefore in this way dispute the hypothesis.

However, the 10 results are quite independent. We may therefore add the values of χ^2 and the degrees of freedom and use these results to test the hypothesis. We thus obtain $\chi^2 = 35.59$ with 10 degrees of freedom. The probability of obtaining this value by chance alone is exceedingly small and hence we must doubt the hypothesis of independence.

We conclude that sex ratio at birth varies with age of the mother.

Testing the difference between two proportions

Questions frequently arise in experimental work and particularly in social surveys as to the interpretation to be placed on different proportions which have been obtained in independent samples. The usual questions are: 'Can the samples have been drawn from the same population?' or 'Are the proportions significantly different?' These questions may be answered by setting out the data in the form of a contingency table and applying the χ^2 test. The same technique may be applied if three or more proportions are given.

Example 11.3

Questions were asked of 400 Italian migrants and 600 English migrants concerning a certain aspect of Australian immigration policy. Three-hundred and twenty-eight of the former and 422 of the latter were in favour. Is there any real difference in the proportions in favour?

Solution

On the assumption that there is no difference in the proportions in favour we obtain the following table of actual and of expected numbers (placed in brackets) in the various categories:

	In favour	Against	Total
Italian	328 (300)	72 (100)	400
English	422 (450)	178 (150)	600
Total	750	250	1000

$$\chi^2 = \frac{(328-300)^2}{300} + \frac{(422-450)^2}{450} + \frac{(72-100)^2}{100} + \frac{(178-150)^2}{150}$$
$$= 17.4$$

There is only one degree of freedom since, when one of the four values within the table is known, the other three may be calculated.

χ^2 tables show that the probability of a value of χ^2_1 of this magnitude occurring by chance is exceedingly small. We conclude that there must be a real difference between the two groups in the proportion in favour.

Example 11.4

The proportion of failures in a class of 100 has changed from 0.42 to 0.5. Is this a significant difference?

Solution

We obtain the following table of actual and expected numbers:

	Pass	Fail	Total
Initial performance	58 (54)	42 (46)	100
Later performance	50 (54)	50 (46)	100
Total	108	92	200

$$\chi^2 = \frac{(58-54)^2}{54} + \frac{(50-54)^2}{54} + \frac{(42-46)^2}{46} + \frac{(50-46)^2}{46}$$
$$= 1.29$$

At the 5% level with one degree of freedom values of χ^2 up to 3.84 may be expected by chance. The value obtained is well below this and hence it is quite likely that the difference is purely due to chance.

Exercises

11.1 The number of persons visiting a certain section of a library during the 10 hours it was open on a certain day were:

$$14, 8, 4, 19, 12, 11, 14, 6, 8, 4$$

Are these figures consistent with the belief that visits are uniformly spread over the 10 hours?

11.2 Four-hundred persons suffering from a nervous complaint were given a bottle of medicine to take in specific doses every three hours. Two-hundred of the bottles contained pure water, the other 200 a tasteless mixture for which substantial claims had been made. The patients reported as follows:

	Better	Worse	No change
Water	110	40	50
Mixture	125	30	45

Are the claims for the mixture justified?

11.3 The colour of hair and colour of eye of 6,800 males are set out in the following table:

Eye colour	Hair colour				Totals
	Fair	Brown	Black	Red	
Blue	1,768	807	189	47	2,811
Grey or Green	946	1,387	746	53	3,132
Brown	115	438	288	16	857
Totals	2,829	2,632	1,223	116	6,800

Are hair and eye colour associated?

11.4 The following results of inoculation against cholera were reported by Greenwood and Yule[2]:

[2] M. Greenwood and G. U. Yule, 'The statistics of anti-typhoid and anti-cholera inoculations, and the interpretations of such statistics in general', *Proceedings of the Royal Society of Medicine*, vol. 8, 1915, p.113

	Not attacked	Attacked	Totals
Inoculated	276	3	279
Not inoculated	473	66	539
Totals	749	69	818

Does inoculation give immunity?

11.5 Six separate experiments were carried out to test a given hypothesis. The results of the experiments gave the following values of χ^2:

χ^2	Degrees of freedom
3.26	1
2.68	1
3.05	1
3.46	1
2.25	1
3.09	1

Is this hypothesis tenable?

11.6 The expected distribution of MN blood types in an Indian tribe are

M	MN	N
.830	.162	.008

The observed frequencies were 375, 77 and 6 respectively. Is the departure from expectation significant?

11.7 The following data was obtained by T. L. Woo[3]:

	Left-eyed	Ambiocular	Right-eyed	Totals
Left-handed	34	62	28	124
Ambidextrous	27	28	20	75
Right-handed	57	105	52	214
Totals	118	195	100	413

Is there any connection between laterality of hand and of eye?

[3] T. L. Woo, 'Dextrality and sinistrality of hand and eye', *Biometrika*, vol. 20A, 1928, p.79

11.8 A motor vehicle manufacturer is considering adding a particular new feature to future models of his car. He shows it to 300 women and 400 men. Of the women, 216, and of the men, 318 were in favour. Is there a real difference of opinion between the sexes?

11.9 Of 200 doctors' sons 17% followed the same occupation as their father. Of 100 lawyers' sons 19% followed their father's occupation. Is it likely that the true proportion of sons following their fathers' occupation is the same for both professions?

12

Sign Tests

Tests when the population distribution is not known

We are often required to compare observed values with those expected on some hypothesis; very often the probability that the observed value *exceeds* the expected is equal to the probability that the observed value *is less than* the expected. If we write a $+$ sign in the former case and a $-$ sign in the latter, it is easy to design a test to see whether a significant departure from the results expected on the hypothesis has occurred. Although such tests are really no more than another application of previous theory they are specifically called 'sign tests'. They may make use of only a part of the information in the original data, and are thus not usually very powerful; they do however have the advantage that no assumption need be made about the population distribution, and are thus amongst a large and useful class of tests often described as 'distribution-free' or (more loosely) 'nonparametric'.

Example 12.1

In an endeavour to maintain consistent standards, a large university follows the practice of passing each year 80% of the students in first year mathematics. A research worker decided to investigate the claim that first year students under age 18 are 'superior' in mathematics. He studied the results of the past 20 years and recorded a $+$ sign for any year in which more than 80% of students *under* age 18 had passed and a $-$ sign in the opposite event. He obtained 17 $+$ signs and 3 $-$ signs. Is the claim justified?

Solution

1. Hypothesis Assume that the claim is not justified and that students under age 18 have the same success rate as students generally.

2. Deduction On this assumption the chances of a + sign and of a — sign are both about $\frac{1}{2}$. We require the chance of obtaining 17 or more from a binomial distribution with $n = 20$ and $\pi = \frac{1}{2}$. Although binomial tables should be used for such small values of n, we shall use the normal approximation to the binomial.

$$\mu = \quad 20 \times \tfrac{1}{2} \quad = 10$$

and
$$\sigma = \sqrt{20 \times \tfrac{1}{2} \times \tfrac{1}{2}} = 2.24$$

The probability required is that of exceeding a standard score of $\dfrac{6.5}{2.24}$ or 2.90. This probability equals 0.0019.

3. Conclusion It is most unlikely that the difference in performance is solely due to chance. We may reasonably conclude on this evidence that students under age 18 are 'superior' in mathematics.

Example 12.2

Two sleep-producing drugs were each tried on 10 patients in the manner described in *Example 10.6* on page 107, and the hours of sleep produced by drug A and by drug B were as follows:

| Drug A | 10.6 | 9.4 | 11.2 | 12.3 | 10.0 | 10.4 | 11.2 | 12.1 | 13.0 | 11.2 |
| Drug B | 10.0 | 12.6 | 10.3 | 10.6 | 9.8 | 9.9 | 8.9 | 11.3 | 8.1 | 10.1 |

Is drug A more effective?
Solution
It is clear from the wording of the question that each of the 10 patients has been administered drug A and drug B, and hence we may apply a paired t test to the differences of the above figures, that is to the extra hours of sleep produced by drug A. These differences are:

$$+0.6, \ -3.2, \ +0.9, \ +1.7, \ +0.2, \ +0.5, \ +2.3, \ +0.8, \ +4.9, \ +1.1$$

a. We assume that there is no difference between the effects of the two drugs. If we can assume that the distribution of hours of sleep produced is near normal we can apply the t test, as in the example referred to on page 107.

$$\bar{x} = 0.98 \qquad\qquad s^2 = 3.614$$

$$\therefore t = \frac{0.98}{\sqrt{3.614}} \sqrt{9} \quad = 1.55$$

At the 5% level of significance with a one-tailed test and nine degrees of freedom, values of t up to 1.83 may be expected. (We have used a one-tailed test because of the definite suggestion that drug A is the more effective.) On this test we therefore cannot refute the hypothesis that the drugs are equally effective.

b. We could also test whether drug A is more effective than drug B by using a sign test, for on our assumption that the drugs are equally effective we would expect drug B to produce more hours of sleep than drug A just as frequently as the reverse situation. That is, the signs of the above differences should be $+$ and $-$ with equal probability. We therefore perform what might be called a paired sign test. In 9 cases out of the 10 we have $+$ signs. We will calculate the probability of 9 or more $+$ signs using the normal approximation to the binomial, although it is hardly justified with such small numbers.

$$\mu = \quad 10 \times \tfrac{1}{2} \qquad\qquad = 5$$

and $\quad \sigma = \sqrt{10 \times \tfrac{1}{2} \times \tfrac{1}{2}} \quad = 1.58$

The probability required is the area under the normal curve beyond $8\tfrac{1}{2}$, which equals the probability of a standard score exceeding $\dfrac{8\tfrac{1}{2} - 5}{1.58}$ or 2.22, which is 0.0132. Thus the probability of drug A being more effective 9 times out of 10 purely as a result of chance is small. We therefore conclude that drug A is more effective.

This is one of those rather unusual cases where the sign test produces a significant result where the usually more powerful t test does not.

Test for the population median

The median has the property that one half of the individuals in the population have values greater than it and one half less, irrespective of the nature of the distribution. If therefore we select individuals at random from a population with a given median, the chance of obtaining a value exceeding the median, and the chance of obtaining a value less than the median are both one half. We can use this property to devise a sign test to test whether a random sample can have come from a population with a given median. This test is clearly valid whatever the population distribution.

Example 12.3

Suppose we are told that the ages of students registering this year for a master's degree are

$$21, 35, 20, 22, 23, 36, 20,$$
$$21, 27, 20, 23, 31, 24, 21, 22.$$

From past experience we know that the median age of persons registering is 25. Are students now enrolling at younger ages?

Solution

We could not use a t test because it is generally known that ages of students registering for higher degrees do not follow a normal distribution. There is a peak in the early twenties and a gradual fall off with increasing age. We cannot use the mean and the Central Limit Theorem approach discussed on page 101 because the sample is too small. We can however use the sign test.

1. Hypothesis We assume that there is no change in the median age of enrolling.

2. Deduction Writing a $+$ when the age is above 25 and a $-$ when the age is below 25 we obtain the following:

$$-\ +\ -\ -\ -\ +\ -\ -\ +\ -\ -\ +\ -\ -\ -$$

There are thus 4 cases above 25 out of 15. This is like getting 4 heads out of 15 tosses and we shall work out approximately the chance of getting 4 or less heads using the fact that the binomial distribution approaches the normal.

$$\mu = \quad 15 \times \tfrac{1}{2} \qquad\qquad = 7.5$$
$$\text{and} \quad \sigma = \sqrt{15 \times \tfrac{1}{2} \times \tfrac{1}{2}} \qquad = 1.94$$

We require the chance of obtaining 4 or less, which corresponds to the probability of a value of less than $4\tfrac{1}{2}$ in the appropriate normal distribution. This equals the probability of obtaining a standard score of less than $\dfrac{-3}{1.94}$ or -1.55, i.e. approximately 0.06. (We used a one-tailed test because of the suggestion that students might be enrolling at *younger* ages.)

3. Conclusion About 6 times in 100 we would expect to obtain, from a population with median age 25, a sample as unusual or more unusual than this one. This is not unlikely enough to invalidate our hypothesis. We cannot therefore on this evidence conclude that there has been a lowering of the median age of enrolling.

Several samples from an unknown population

Given details of two samples, we may like to know if they could reasonably have been drawn from the same population. If the samples were large, we could invoke the Central Limit Theorem and test the difference between the sample means using the method described on page 101 . In the case of small samples, if we know that the population was normally or nearly normally distributed then we could use the t test to test the differences between the sample means using the method described on page 105. Neither of these methods deals with the case of small samples drawn from a population whose distribution is unknown or which is known *not* to be normally distributed. It is possible to apply a form of sign test which may give some indication as to whether it is reasonable to assume that the two samples were drawn from the same population. This test, called by some writers 'the median test', may also be applied where there are several samples.

The steps involved in this test are as follows:
1. Combine all the samples to form an artificial population and assume that the samples are drawn at random from this population;
2. Arrange this artificial population (i.e. the combined samples) in the form of an array;
3. Determine the median of this artificial population;
4. For each sample, record a $+$ sign for each value which exceeds the median found in   *3* , and a $-$ sign for each value below that median and present the number of $+$ and $-$ signs for each sample in the form of a contingency table;
5. Apply a χ^2 test to this contingency table. The number of degrees of freedom is $n - 1$, where n is the number of samples, since if the number of $+$ signs for $n - 1$ samples is known, the number of $+$ signs for the remaining sample and the number of $-$ signs for all the samples can be determined from the number of items in each sample.
The test thus treats the combined samples as a population whose median can therefore be determined. If n samples of the given sizes were drawn at random from this artificial population the expected number of $+$ signs and the expected number of $-$ signs for each sample would each equal half the number of items in that sample. The χ^2 test measures the extent of the departure of the actual number of $+$ and $-$ signs from these expected numbers. If this value of χ^2 is significantly large, we conclude that the n samples have not been drawn

at random from this one 'combined-sample' population. Unless we wish to dispute this assumption of randomness we would thus presume that they have been drawn from two or more different populations.

Example 12.4

L. C. Freeman and A. P. Merriam[1], studying the musical style of the Ketu cult of Brazil and the Rada cult of Trinidad, obtained the following proportions of minor thirds in two samples of twenty songs from each cult:

Rada Type	Ketu Type
.529	.211
.281	.327
.412	.204
.541	.115
.574	.199
.399	.278
.416	.023
.350	.100
.203	.484
.471	.121
.359	.363
.394	.108
.550	.081
.332	.136
.423	.280
.440	.411
.385	.170
.440	.189
.414	.152
.249	.220

Is there any difference between the two cults in musical style?
Solution
1. Combine the two samples to form an artificial population and assume that the two samples have been drawn at random from this population.

[1] L. C. Freeman and A. P. Merriam, 'Statistical classification in anthropology: an application to ethnomusicology', *American Anthropologist*, vol. 58, 1956, p.464

2. The median of this population lies between .327 and .332.

3. The number of cases in each cult which are less than and the number which are greater than this median are set out in the following table:

Cult	Less than median	More than median	Total
Ketu	17	3	20
Rada	3	17	20
Total	20	20	40

The expected numbers less than and greater than the median, are both 10, in the case of each cult.

4. Hence
$$\chi^2 = \frac{(17-10)^2}{10} + \frac{(3-10)^2}{10} + \frac{(3-10)^2}{10} + \frac{(17-10)^2}{10}$$

$$= 19.6 \text{ with 1 degree of freedom.}$$

5. The chance of obtaining a value of χ^2 of this size or greater with one degree of freedom is exceedingly small. It is most unlikely that the two samples have been drawn from a population made up of the two samples combined.

6. We conclude that there are differences in musical style between the two cults.

Example 12.5

A consumer protection association is investigating the durability of four brands of electric radiator. They therefore subject several samples of each brand to conditions simulating regular use, and observe the number of months taken for a major fault to develop. The results of their investigations are as follows:

Radiator type	Months to failure							
Brand A (soldered connections)	5	8	12	15	16	19	25	33
Brand B (,, ,,)	5	6	8	10	11	13	15	16
Brand C (integral construction)	18	19	20	24	28	30		
Brand D (,, ,,)	12	14	19	20	21	23	25	32

Is there any definite difference among the various brands?

Solution

1. Combine the four samples to form an artificial population and assume that the four samples are selected at random from this population.

2. The median of this population is 17.

3. Assigning a $+$ to values above the median and a $-$ to values below the median we obtain the following table of signs:

Brand A	$-$	$-$	$-$	$-$	$-$	$+$	$+$	$+$
Brand B	$-$	$-$	$-$	$-$	$-$	$-$	$-$	$-$
Brand C	$+$	$+$	$+$	$+$	$+$	$+$		
Brand D	$-$	$-$	$+$	$+$	$+$	$+$	$+$	$+$

It looks reasonably certain that the sample of brand B would not have been drawn from a population of median 17 as the signs are all negative. The chance of this would be $(\frac{1}{2})^8 = 1$ in 256.

Similarly the chance of the sample of brand C being drawn from such a population would be $(\frac{1}{2})^6$ of 1 in 64.

The distribution of signs is set out in the following contingency table:

Brand	$+$	$-$	Total
A	3	5	8
B	0	8	8
C	6	0	6
D	6	2	8
Total	15	15	30

Care would need to be used in analysing this table because of the low *expected* frequencies which occur. We note that in the samples of brands A and B, $-$ signs predominate and in those of brands C and D $+$ signs predominate. We might therefore like to test whether the samples of brands A and B could have come from the same population as the samples of brands C and D, i.e. to test whether the brands with soldered connections are really different from those of integral construction. If we combine the samples of brands A and B and also combine the samples of brands C and D we obtain the distribution of signs in the following table:

Samples	+	—	Totals
Brands A and B combined	3	13	16
Brands C and D combined	12	2	14
Total	15	15	30

4. The expected number of + and — signs are each 8 for the samples of brands A and B combined, and 7 for the samples of brands C and D combined.

$$\text{Hence} \quad \chi^2 = \frac{(3-8)^2}{8} + \frac{(13-8)^2}{8} + \frac{(12-7)^2}{7} + \frac{(2-7)^2}{7}$$

$$= 13.39 \text{ with 1 degree of freedom}$$

5. The chance of obtaining a value of χ^2 of this size or greater with one degree of freedom is very small. It is most unlikely that the samples have been drawn from a population made up of the samples combined.
6. We conclude that the samples of brands A and B have not been drawn from the same population as the samples of brands C and D. We conclude that radiators with soldered connections have a different effective life from those of integral construction.

Sample values equal to the median

Quite often one, two or more of the individual sample values are equal to the median. In *Example 12.3* we could have had students registering who were age 25. The method of dealing with such cases is to divide them evenly between the 'above median' and 'below median' groups. If there are an odd number of such cases, they should be divided equally and the last case ignored (see Exercise *12.1*).

Visual inspection of signs

In situations where it is appropriate to use a sign test it may not be necessary actually to perform the calculations; a quick inspection of the signs may of itself be sufficient to give an indication as to whether a significant departure from the hypothesis has occurred.

Example 12.6

G. N. Pollard[2], investigated whether the sex ratio at birth varied with

[2] G. N. Pollard, 'Factors influencing the sex ratio at birth in Australia 1902-65', *Journal of Biosocial Science*, vol. 1, 1969, p.125

age of mother. He analysed Australian data from 1914 to 1963 (in 5 year groups) for various ages of mother and in only one 5 year period could he detect, using the χ^2 test, any significant difference between the actual number of male births in the various categories and those expected on the hypothesis of constant sex ratio.

However if a $+$ sign is used to indicate that the actual number of male births exceeded expected (on the assumption of a sex ratio not varying with age), and a $-$ sign the reverse, then his results were as follows:

Age of mother	Years									
	1914–1917	1918–1922	1923–1927	1928–1933	1934–1938	1939–1943	1944–1948	1949–1953	1954–1958	1959–1963
−19	+	+	+	−	+	+	+	+	+	+
20–24	+	+	+	+	+	+	+	+	+	+
25–29	+	−	+	−	+	+	−	−	+	+
30–34	+	+	+	+	−	+	−	+	−	−
35–39	−	−	−	−	−	−	−	−	−	−
40–	−	−	−	−	−	+	−	+	−	−

Can we conclude that the sex ratio of births varies with mother's age?

Solution

Assume the sex ratio at birth does not vary with mother's age. Then if on this hypothesis we calculate the expected number of male births, the actual number is equally likely to be more or less than the expected. That is, on our hypothesis, $+$ and $-$ signs are equally likely in the above table.

The actual distribution of signs (with the expected numbers in brackets) is as follows:

Age of mother	Number of		Total
	+ signs	− signs	
Under 25	19 (10)	1 (10)	20
25-34	12 (10)	8 (10)	20
35 and over	2 (10)	18 (10)	20
Total	33	27	60

Hence $\chi^2 = \dfrac{(19 - 10)^2}{10} + \dfrac{(1 - 10)^2}{10} + \ldots + \dfrac{(18 - 10)^2}{10}$

$\qquad\qquad = 29.8$

There are 3 degrees of freedom, since the total number of $+$ signs and the total number of $-$ signs are not fixed.

This is too large a value of χ_3^2 to have occurred by chance and we conclude that sex ratio at birth varies with mother's age.

This conclusion could have been deduced from a visual inspection of the table. Consider ages under 25, i.e. above the top broken line. There are 19 $+$ signs and 1 $-$ sign. The probability of this happening due to chance alone is so small that we need not make the calculation.

Consider ages 35 and over, i.e. below the lower broken line. Here there are 18 $-$ signs and 2 $+$ signs. Here again the probability of this happening due to chance alone is very small.

In the age group 25 to under 35 there are 12 $+$ signs and 8 $-$ signs, indicating no departure from hypothesis.

We thus conclude that our hypothesis of constant sex ratio with age of mother is wrong and that masculinity of births is higher than average for mothers under 25 and lower than average for mothers 35 and over.

It should be emphasised that for a problem such as this the usual approach would be to use the figures for the actual and expected births and apply the χ^2 test. This approach was in fact adopted but the differences between actual and expected were so small that the χ^2 test did not indicate a significant discrepancy. The small deviations were however consistently one way and the sign test established a significant departure from the hypothesis of constant sex ratio. It is unusual for a sign test to produce a significant result when the χ^2 test does not. This example shows the desirability of applying to the results of an experiment two or more tests when these are available.

Exercises

12.1 The ages of the 32 players accepted this year for a men's singles tennis title were:

$$25, 17, 19, 20, 18, 19, 28, 29, 23, 22, 19, 22,$$
$$21, 23, 18, 22, 20, 21, 35, 18, 37, 19, 22, 23,$$
$$28, 21, 31, 22, 20, 26, 18, 29.$$

For some years the median age has been 23. Are the tennis players getting younger?

12.2 A tutor in introductory statistics who had been criticised for his unorthodox methods of teaching claimed that they were more effective. To prove his case he arranged the students in two groups, selecting them in pairs with the same background, academic record, sex etc. and placed one in his own class (class A) and the other in class B, taken by a colleague. The final examination marks were:

Class	Marks																
A	69	68	65	74	65	64	60	75	75	73	79	63	75	82	81	69	80
B	63	65	72	69	61	68	66	71	74	75	72	67	72	76	68	69	72

Use the sign test to determine whether his claims are justified.

12.3 In his investigation into the sex ratio at birth in Australia, G. N. Pollard[3], compared the actual number of male births with the expected (assuming no variation with parent's age) for every combination of mother's and father's age recorded in individual years. Writing + when actual exceeded expected and — when actual were less than expected he thus obtained a very large two-way table of + and — signs. Adding up the number of signs for all ages of father for quinquennial age groups of mother he obtained the following table:

Mother's age	Number of +	Number of —
— 19	13	8
20 — 24	59	45
25 — 29	39	52
30 — 34	34	45
35 — 40	30	41

[3] G. N. Pollard, 'Factors influencing the sex ratio at birth in Australia 1902-65', *Journal of Biosocial Science* , vol. 1, 1969, p.125

Can we conclude that masculinity of birth varies with the mother's age?

12.4 A scientist wished to determine whether a certain characteristic of the blood of rats varies with age. He had no knowledge of the way in which the characteristic was distributed. For 40 rats of the same species and kept under the same conditions he obtained the following results:

Age Group	Measurement									
1	19	21	18	26	26	22	20	27	19	22
2	24	17	22	20	26	21	28	20	27	19
3	21	29	26	22	20	25	19	21	28	30
4	26	26	25	24	29	22	30	27	21	28

Test the hypothesis that these are four samples from the same population.

12.5 The same scientist measured the same characteristic in 20 rats both under normal conditions and later after considerable exercise. The results obtained were

Normal	19	24	22	24	20	25	24	28	26	24
After Exercise	23	28	23	20	26	30	27	24	27	27
Normal	19	18	21	28	23	20	25	25	26	19
After Exercise	23	21	25	22	25	26	29	19	31	24

Does exercise have a significant effect on the characteristic?

13

Product Moment Correlation

So far we have been concerned only with a single variable and its frequency distribution. For example we studied earlier the number of calls made by insurance men in a week. Many problems in statistical work are bivariate problems i.e. they are concerned with two variables. We might have been given, not only the number of calls, but also the weekly earnings of each of the men. In such a situation we are often interested to see whether there is any inter-relation between the two. Are the men with high weekly earnings usually those who make many calls and vice versa? This is a correlation problem.

Example 13.1

Let us compare the weekly earnings of the first 20 insurance men with the number of calls they made during that week. Here are the figures:

Calls	184	172	171	125	173	144	207	174	155	152
Earnings ($)	114	94	104	58	100	70	120	108	92	94
Calls	117	135	122	161	204	149	140	138	200	176
Earnings ($)	60	80	72	106	114	76	76	64	118	92

Is there any relationship between calls and earnings?

It is difficult to tell simply by inspection, but we could obtain some idea if we were to present the data in the form of a graph, as in *Fig. 13.1*. This type of graph is called a scattergraph or a scatter diagram. This graph makes clear what was not readily apparent from the figures, that the higher earnings are made by those men who make the larger number of calls and vice versa.

If we had, for convenience, measured the number of calls from an arbitrary origin of 160 (i.e. recorded a person who made 180 calls as

20, one who made 150 calls as — 10, and so on) and if we had measured earnings from say $80 (i.e. recorded a person who earned $90 as 10, one who earned $60 as — 20, and so on) our graph — that is, the relative position of all the points — would be unchanged. In fact the only change in the scatter diagram would be a change in the figures along the x axis to —60, —40, —20 etc. and along the y axis to —40, —20 etc. A suitable point to select as origin, for reasons which we shall see later, is the arithmetic mean.

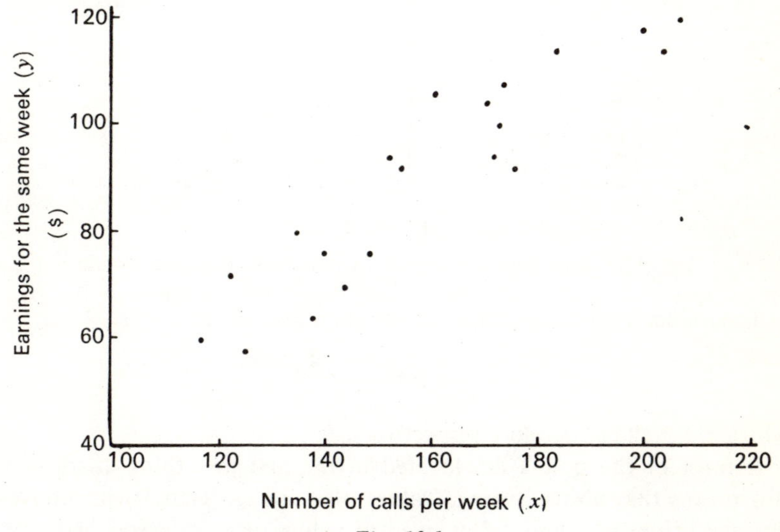

Fig. 13.1

Number of calls per week and earnings for 20 insurance men

Further, we could equally have changed the unit of earnings to cents, pounds or units of $10 and also made changes in the units in which calls are measured, but provided we were still representing the same basic information the relationship between the points — the scatter diagram — would not change. Similarly, if we were studying the relationship between the height and weight of individuals (whether tall people weigh more and vice versa) this relationship will not be affected if we measure in pounds and inches or in grams and centimetres or in any other combination of units. We shall use as our units in this exercise the standard scores for x and y, for if we do so both origin and scale are then independent of the original units. If this is done our scatter diagram appears as in *Fig. 13.2*.

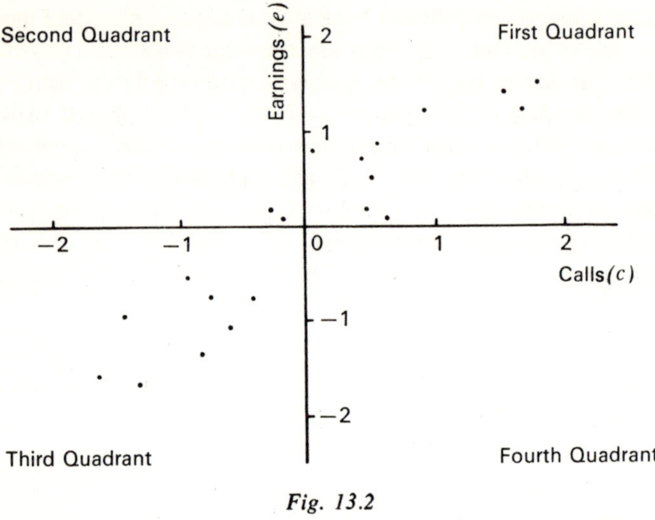

Fig. 13.2

Fig. 13.1 data plotted in standard units with means as origin

The values we have plotted are the standard scores

$$\frac{x - \bar{x}}{s_x} \text{ and } \frac{y - \bar{y}}{s_y}$$

which we shall call c and e respectively.

If most of the points are located in the first and third quadrants, this means that above-average values of c are associated with above-average values of e and below-average values of c are associated with below-average values of e. If most of the points are located in the second and fourth quadrants, this means that the larger values of c are associated with the smaller values of e and smaller values of c are associated with larger values of e.

The correlation coefficient (r)

It would be useful to have some numerical method of measuring the extent to which (say) above-average values of c are associated with above-average values of e. One procedure is to take the average value of $c \times e$ for all points. This is called the product moment correlation coefficient r.

Thus $\qquad r = \frac{1}{n} \Sigma c \times e$

In the first quadrant c and e are both positive and in the third quadrant both are negative. Hence in both of these quadrants the contribution to r is positive. From the second and fourth quadrants the contribution to r is negative. Hence large positive values of r indicate that high values of c correspond with high values of e and vice versa. Large negative values of r indicate that high values of c correspond with small values of e and vice versa.

It can be shown that the highest possible value of r is $+1$ and the lowest possible value -1.

Terms used to describe correlation

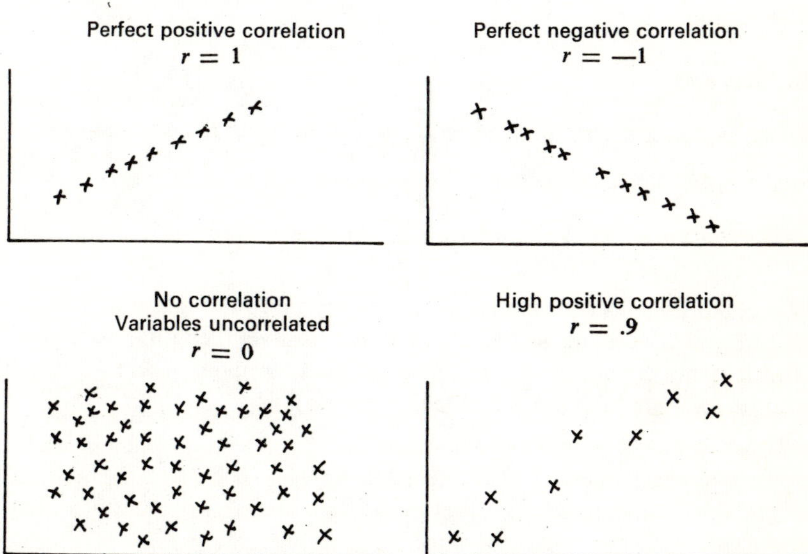

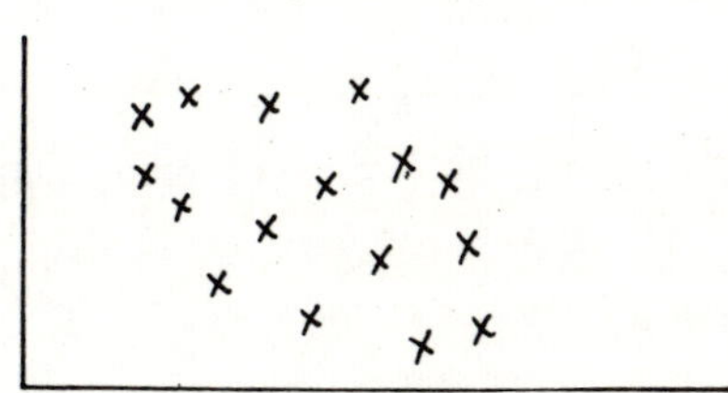

Fig. 13.3

If all the points in a scatter diagram lie on a straight line the correlation is described as perfect. If with perfect correlation positive values of c are associated with positive values of e we have perfect positive correlation and $r = 1$. If with perfect correlation positive values of c are associated with negative values of e we have perfect negative correlation and $r = -1$.

When there is no pattern among the points, i.e. when they are scattered at random, the variables are said to be uncorrelated or there is said to be no correlation and $r = 0$.

If the points approach near to the straight line situation we have high correlation; if they approach the random situation we have low correlation.

Formulae for r

If we express the formula $r = \frac{1}{n} \Sigma c \times e$ in terms of the original values we obtain

$$r = \frac{\Sigma (x - \bar{x})(y - \bar{y})}{n\, s_x\, s_y}$$

We could use this formula to calculate r by first finding $\bar{x}$ and $\bar{y}$, then $x - \bar{x}$ and $y - \bar{y}$ and so on. This however is a very lengthy procedure and it is much shorter to use the equivalent formula (where x and y can be measured from *any* convenient origin)

$$r = \frac{\frac{1}{n} \Sigma xy - \bar{x}\bar{y}}{s_x\, s_y}$$

where, as on page 45, $s_x = \sqrt{\dfrac{\Sigma x^2}{n} - \bar{x}^2}$

and $s_y = \sqrt{\dfrac{\Sigma y^2}{n} - \bar{y}^2}$.

The calculation of r for the 20 insurance men, using this formula, is given in *Table 13.1*.

Calculation of r from a bivariate frequency table

In our previous calculation we were given values of calls and earnings for each individual. When data in respect of a large number of persons

Table 13.1

Calculation of *r* for the example on page 132

x	y	x^2	y^2	xy
184	114	33856	12996	20976
172	94	29584	8836	16168
171	104	29241	10816	17784
125	58	15625	3364	7250
173	100	29929	10000	17300
144	70	20736	4900	10080
207	120	42849	14400	24840
174	108	30276	11664	18792
155	92	24025	8464	14260
152	94	23104	8836	14288
117	60	13689	3600	7020
135	80	18225	6400	10800
122	72	14884	5184	8784
161	106	25921	11236	17066
204	114	41616	12996	23256
149	76	22201	5776	11324
140	76	19600	5776	10640
138	64	19044	4096	8832
200	118	40000	13924	23600
176	92	30976	8464	16192
3199	1812	525381	171728	299252

$$\bar{x} = \frac{3199}{20} = 159.95 \qquad \bar{y} = \frac{1812}{20} = 90.6$$

$$s_x = \sqrt{\frac{525381}{20} - 159.95^2} = 26.18 \qquad s_y = \sqrt{\frac{171728}{20} - 90.6^2} = 19.44$$

$$r = \frac{\frac{1}{20} \times 299252 - 159.95 \times 90.6}{26.18 \times 19.44} = +0.93$$

is being studied this is too cumbersome and it is usual to present the data in summary form. The usual form is that of a bivariate frequency table.

The information contained in *Table 13.2* within the square bounded by the heavy lines is in this form. It indicates the number of our 80 insurance men whose earnings and whose number of calls fell within certain specified groups. The correlation coefficient is often required for data presented in this form.

We showed (*Table 4.4*, page 48) when calculating the standard deviation that the calculation was simplified if we worked from an arbitrary origin and also changed the scale. In calculating *r* also it simplifies our work if we change the origin and the scale of the variables *x* and *y* to new variables *a* and *b* (say) and use the fact that the value of *r* is the same for *x* and *y* as it is for *a* and *b*.

Table 13.2

The calculation of r from a bivariate frequency table

Earnings in the same week $	\ 100– 119	120– 139	140– 159	160– 179	180– 199	200– 219	Totals (fy)	b	fyb	fyb²	bΣfa
	Number of calls per week										
115 – under 125					1	5	6	3	18	54	51
105 – under 115		2	5	3	2		12	2	24	48	34
95 – under 105		3	6	3			12	1	12	12	12
85 – under 95	1	9	5				15	0	0	0	0
75 – under 85	1	2	9	1			13	—1	—13	13	3
65 – under 75	2	7	3				12	—2	—24	48	22
55 – under 65	5	5					10	—3	—30	90	45
Totals (fx)	8	15	26	17	7	7	80		—13	265	167
a	—2	—1	0	1	2	3			=Σfyb	=Σfyb²	=Σfab
fxa	—16	—15	0	17	14	21	21	=Σfxa			
fxa²	32	15	0	17	28	63	155	=Σfxa²			
aΣfb	40	31	0	15	24	57	167	=Σfab			

Hence $s_a^2 = \dfrac{155}{80} - \left(\dfrac{21}{80}\right)^2 = 1.8686 \qquad s_b^2 = \dfrac{265}{80} - \left(\dfrac{-13}{80}\right)^2 = 3.2861$

$$r = \frac{\dfrac{167}{80} - \left(\dfrac{21}{80}\right)\left(\dfrac{-13}{80}\right)}{\sqrt{1.8686} \times \sqrt{3.2861}} = +\,.86$$

N.B. The figures under the heading $b\Sigma fa$ are obtained thus;

$51 = 3(1 \times 2 + 5 \times 3)$ as $b = 3$, and $f = 1$ and 5 for $a = 2$ and 3 respectively

$34 = 2(2 \times 0 + 5 \times 1 + 3 \times 2 + 2 \times 3)$ etc.

Similarly for the figures on the line $a\Sigma fb$. Thus:

$40 = -2(1 \times -1 + 2 \times -2 + 5 \times -3)$.

When the data is grouped we need to modify the above formula for r in the same way as we modified the formula for the standard deviation on page 47, For grouped data if we write f for the frequency within any group, the formula for the correlation coefficient between a and b becomes

$$r = \frac{\frac{1}{n}\,\Sigma f\,ab - ab}{s_a\ s_b}$$

where, as shown in Chapters 3 and 4.

$$\bar{a} = \frac{\Sigma fa}{n} \qquad\qquad \bar{b} = \frac{\Sigma fb}{n}$$

$$s_a = \sqrt{\frac{\Sigma fa^2}{n} - \bar{a}^2} \qquad\qquad s_b = \sqrt{\frac{\Sigma fb^2}{n} - \bar{b}^2}$$

The frequencies of x values are used to calculate $\bar{a}$ and s_a and of y values to find $\bar{b}$ and s_b as indicated in the rectangles lying outside the square containing the data. The method of calculating Σfab is also shown. This is calculated in two ways which serves as a check on the work.

The method of calculation can be followed by working tnrough the calculation attached to our bivariate frequency table in *Table 13.2.*

Spurious correlation

Great care must be taken when interpreting values of r. It does not follow that because there is a high correlation between two variables that a change in one variable will cause a change in the other. Correlation and causation are different concepts. High values can be obtained for r in cases where there is no *direct* connection between the variables. This is termed spurious correlation.

The number of teachers of statistics, the number of taxation officials, etc. and the consumption of liquor could have a correlation of nearly unity because both could have a steady upward trend in good times. We cannot deduce from this any cause and effect relationship.

Considerable experience in the particular field is necessary to enable an experimenter to assess causality aspects when significantly high values of r have been obtained.

Testing hypotheses concerning r

If we select a large number of samples of size n from a bivariate population which has a two dimensional normal distribution in which the variates have a correlation coefficient ρ, and work out values of r for each sample, then the distribution of r can be obtained and we could determine how unlikely it is that any particular value of r would be obtained from a random sample from this population. The distribution of r is, however, very complicated and depends on both ρ and n. It is not therefore generally used.

It has been found, however, that if, for each of these many samples, we calculate from each value of r a value of w where

$$w = \tfrac{1}{2} \log_e \frac{1 + r}{1 - r}$$

then the values of w follow a very nearly normal distribution with mean

$$\tfrac{1}{2} \log_e \frac{1 + \rho}{1 - \rho}$$

and standard deviation

$$\frac{1}{\sqrt{n - 3}}$$

This remarkable result enables us to use normal distribution tables to test the hypothesis that a particular sample value of r could have been obtained by chance from a bivariate normal population with a given correlation coefficient ρ.

Fortunately it is not necessary to calculate values of w using this formula because values of w corresponding to given values of r have been tabulated and are set out in *Appendix II* (page 259).

The usual problem is to determine whether a random sample with a particular value of r could reasonably have come from a population in which there is no relationship between the two variables i.e. where $\rho = 0$. For this problem we determine the value of w from the tables and test whether it is likely to have come from a population which is normally distributed with mean zero and standard deviation $\dfrac{1}{\sqrt{n - 3}}$.

Example 13.2

Could the sample correlation coefficient of 0.86 between the number of calls and the earnings in a week of 80 insurance men have occurred if there was no real relationship between these variables?

Solution

1. Assume the variables are uncorrelated, i.e. $\rho = 0$. Assume also that they both are normally distributed.

2. The sample value of $r = .86$ and the corresponding value of w (from the tables) is 1.293.

3. On our hypothesis values of w should be normally distributed with mean 0 and SD $\dfrac{1}{\sqrt{77}} = 0.114$.

4. This sample therefore has a value of w which is $\dfrac{1.293}{0.114}$ or 11.3 standard deviations above the mean. It is very unlikely that a standard score as large as this would happen by chance. Thus we must doubt our assumption that $\rho = 0$ and conclude that a significant non-zero correlation has been confirmed.

Example 13.3

The total annual rainfall and the total hours of sunshine in a certain city were as follows:

Year	Rainfall (mm)	Sunshine (hrs.)
1951	245	2096
1952	202	2342
1953	257	2157
1954	206	2074
1955	227	2054
1956	221	2256
1957	230	2067
1958	215	2312
1959	232	2050
1960	298	2111
1961	282	2135
1962	192	1976
1963	252	1875
1964	225	1998
1965	216	1924
1966	209	2246
1967	195	2040
1968	242	1989
1969	262	1842

a. Calculate the coefficient of correlation.

b. Test the hypothesis that there is no correlation between annual rainfall and annual hours of sunshine in this city.

Solution

a. Using as arbitrary origins, rainfall $= 200$ mm and sunshine $= 2000$ hrs. and using x and y respectively for values measured from these arbitrary origins we obtain

$$\Sigma x = 608 \qquad \Sigma y = 1544 \qquad \Sigma xy = 35651$$

$$\Sigma x^2 = 34324 \qquad \Sigma y^2 = 468818 \qquad n = 19$$

Hence $\quad \bar{x} = \dfrac{608}{19} = 32.0 \qquad \bar{y} = \dfrac{1544}{19} = 81.26$

$$s_x = \sqrt{\dfrac{34324}{19} - 32.0^2} = 27.97$$

$$s_y = \sqrt{\dfrac{468818}{19} - 81.26^2} = 134.4$$

giving $\quad r = \dfrac{\dfrac{35651}{19} - 32.0 \times 81.26}{27.97 \times 134.4} = -0.19$

b. Assume that this is a sample from a bivariate normal population in which $\rho = 0$

The value of w corresponding to a value of r of -0.19 is (from the tables) -0.192. On our hypothesis values of w should be normally distributed about mean zero with SD $= \dfrac{1}{\sqrt{19 - 3}} = 0.25$. The sample therefore has a value of w which is $\dfrac{0.192}{0.25}$ or 0.77 standard deviations below the mean. A deviation of this amount or greater would occur quite frequently by chance and hence from this data there is no reason to dispute the hypothesis of no correlation.

Note. We were not given any information as to whether data such as this is approximately normally distributed. If the data had produced a significantly large correlation coefficient, before coming to a firm conclusion we would need to make sure that the normality assumption was a reasonable one.

Exercises

13.1 Calculate the product moment correlation coefficient between the following figures:

$$x \quad 2 \quad 4 \quad 6 \quad 8 \quad 10$$
$$y \quad 1 \quad 3 \quad 5 \quad 7 \quad 9$$

13.2 Calculate the product moment correlation coefficient between the following figures:

$$x \quad 2 \quad 4 \quad 6 \quad 8 \quad 10$$
$$y \quad 9 \quad 7 \quad 5 \quad 3 \quad 1$$

13.3 Calculate the product moment correlation coefficient between the following figures:

$$x \quad 1 \quad 2 \quad 3 \quad 4 \quad 5$$
$$y \quad 5 \quad 5 \quad 0 \quad 5 \quad 5$$

13.4 The marks obtained by 19 students in an english test and a history test were:

English	6 7 6 8 5 6 4 4 9 5 8 7 6 5 9 8 4 7 8
History	4 9 5 6 7 6 4 7 7 4 9 8 6 8 5 8 6 6 8

Plot the scattergram, determine the correlation coefficient and test its significance.

13.5 The following table gives the relative ages of brides and bridegrooms for a small sample of 79 couples. Calculate the value of r and test the hypothesis that the value of ρ in the population from which the sample is drawn is 0.8.

Age of bridegrooms	Age of brides					Total
	under 20	20-4	25-9	30-4	35-	
under 20	4					4
20-4	14	21	1			36
25-9	3	12	4	1		20
30-4	1	3	2	1	1	8
35-		1	1	2	7	11
Total	22	37	8	4	8	79

13.6 The following figures give the number of marriages registered in Australia for each of the 10 years from 1960 to 1969 and the production of brown coal for the same years:

Marriages (,000)

75.4   76.4   77.6   80.7   82.4   90.3   95.1   98.0 102.5 109.6

Production of brown coal (m. tons)

14.1   15.7   16.8   17.8   18.7   19.5   21.6   22.5   23.1   23.1

a. Calculate the correlation coefficient between the two sets of figures.
b. Would you agree that an increase in the number of marriages results in an increase in coal production? Give reasons for your answer.

14

Rank Correlation

Data is often presented, not in the form of actual measurements as in the previous chapter, but in rank order. School children may be listed in order of height or weight, participants in a beauty contest may be ranked by several judges, employees may be ranked by supervisors on their job performance. Because of difficulties in measuring psychological variables the method of ranking is frequently adopted in psychology. We frequently have *two* sets of rankings, for example, the ranking of certain students in english and in history, or the ranking of school children in height and in weight, and we may wish to determine if there is any relationship — any correlation — between the two rankings. In other words, we wish to answer such questions as 'Are the top students in history also the top students in english? or the reverse? or is there no connection?' In Chapter 13 we were dealing with similar questions but there we had the actual marks. Here we will be concerned with answering these same questions given only the rankings.

Also in Chapter 13, even though we had the actual values of the observations it was necessary to assume that the population from which the sample was drawn was a bivariate normal population. If there is no justification for this assumption then that test is not valid. We can however rank the data from the actual values of the observations and use the technique of this chapter. The rank correlation test which we shall describe is one of a number of such tests which are extremely valuable as equivalents of ordinary correlation in situations where we know nothing about the population distribution.

Spearman's coefficient of rank correlation r_s

If the rankings of n students in english and in history are

$$\begin{array}{lccccccc} \text{English} & 1 & 2 & 3 & 4 & . & . & . & n \\ \text{History} & 1 & 2 & 3 & 4 & . & . & . & n \end{array}$$

they are in complete agreement, i.e. they are perfectly correlated.

If the rankings were

| English | 1 | 2 | 3 | 4 | . | . | . | n |
| History | n | $n-1$ | $n-2$ | $n-3$ | . | . | . | 1 |

they are in complete disagreement, i.e. they are perfectly negatively correlated.

C. Spearman[1], suggested measuring the degree of correlation by finding the differences d between the corresponding rankings of individuals, squaring the figures so obtained and adding. He thus based his measurement of correlation on Σd^2.

In the first case the differences are all zero and hence Σd^2 has its minimum value of zero. In the second case, Σd^2 has its maximum value of

$$(n-1)^2 + (n-3)^2 + (n-5)^2 + \ldots + (n-2n+1)^2$$

which can be shown to be equal to $\dfrac{n^3 - n.}{3}$

Spearman therefore suggested as a coefficient of rank correlation

$$r_s = 1 - \frac{6 \Sigma d^2}{n^3 - n}$$

because this has the properties

a. $r_s = +1$ for perfect correlation (i.e. when $\Sigma d^2 = 0$)

b. $r_s = -1$ for perfect negative correlation (i.e. when $\Sigma d^2 = \dfrac{n^3 - n}{3}$)

and for all other situations r_s lies between -1 and $+1$.

The subscript s in r_s stands for 'Spearman' to distinguish this coefficient from the product moment correlation coefficient described in Chapter 13.

Example 14.1

The sports editors of two daily newspapers, just before an annual regatta, forecast the placing of the eight crews in the race as follows:

	Crew							
	A	B	C	D	E	F	G	H
First editor	6	4	2	1	3	7	8	5
Second editor	4	1	6	3	5	8	7	2

[1]   C. Spearman, 'The proof and measurement of association between two things', *American Journal of Psychology*, vol. 15, 1904, p.72

Are the editors in substantial agreement?

Solution
The differences between the ranks are

$$+2 \quad +3 \quad -4 \quad -2 \quad -2 \quad -1 \quad +1 \quad +3$$

$\Sigma d = 0$ as it should.

$$\Sigma d^2 = 4 + 9 + \ldots + 9 = 48$$

$$\therefore r_s = 1 - \frac{6}{504} \times 48 = 0.429.$$

Since the value of r_s is positive but well below $+1$ it indicates some measure of agreement between the editors but not a very strong one.

Example 14.2

In the previous example there was no information about the 'real' order and r_s simply gave some indication of the extent of agreement between the editors. However, after the race had been run the real order became known. The value of r_s between each editor's forecast and the actual placings measures the extent to which he is a good judge.

The result of the race was

			Crew				
A	B	C	D	E	F	G	H
6	5	1	3	2	4	8	7

Were the editors good judges of this kind of prowess?

Solution
a. First editor
The differences in rankings are

$$0 \quad -1 \quad +1 \quad -2 \quad +1 \quad +3 \quad 0 \quad -2$$

$\Sigma d = 0$

$\Sigma d^2 = 20$

$$r_s = 1 - \frac{6}{504} \times 20 = 0.762$$

He is a good judge.
b. Second editor
The differences in rankings are

$$-2 \quad -4 \quad +5 \quad 0 \quad +3 \quad +4 \quad -1 \quad -5$$

$\Sigma d = 0$

$\Sigma d^2 = 96$

$$r_s = 1 - \frac{6}{504} \times 98 = -0.143$$

His was a poor forecast. He may well have done better if he had selected the rankings by chance.

Tied ranks

Where tied ranks occur they are usually replaced by the average of the ranks they replace. Thus if three persons are placed fourth, they fill the positions 4, 5 and 6 and they should each be given the number 5. Two persons placed third should each be given the number $3\frac{1}{2}$.

Testing a value of r_s

If two sets of rankings are written down purely at random it is not unusual to obtain quite large positive values of r_s between them. It is possible to work out the probability, if rankings are written down purely at random, that a given value of r_s will be exceeded by chance. It is therefore usual to make a null hypothesis of zero correlation and use published tables to determine whether the value of r_s obtained is likely to have occurred by chance. Fairly large values of r_s are required to reject the hypothesis of zero correlation, e.g., for $n = 10$ we require a value of r_s of .564 or greater at the 5% level.

Of the three examples just given only in the case of the first editor in *Example 14.2* is doubt cast on the hypothesis of zero correlation.

A table giving significant values of r_s is given in the *Appendix* (page 260).

Example 14.3

In *Example 13.3* of Chapter 13 (page 141) certain rainfall and sunshine data was given. Some doubt was expressed about the justification for the normality assumptions made there and hence a rank correlation approach is appropriate. The data for the various years expressed in rank form is as follows:

Rainfall	14	3	16	4	10	8	11	6	12	19	18	1	15	9	7	5	2	13	17
Sunshine	12	19	15	11	9	17	10	18	8	13	14	4	2	6	3	16	7	5	1

a. Calculate the coefficient of rank correlation.

b. Test the hypothesis of zero correlation.

Solution

a. The differences between the ranks are

$$+2 \quad -16 \quad +1 \quad -7 \quad +1 \quad -9 \quad +1 \quad -12 \quad +4 \quad +6$$
$$+4 \quad -3 \quad +13 \quad +3 \quad +4 \quad -11 \quad -5 \quad +8 \quad +16$$

$\Sigma d = 0$

$\Sigma d^2 = 1270$

$$r_s = 1 - \frac{6}{6840} \times 1270 = -0.11$$

b. From the rank correlation table (*Appendix*, page 260) for a sample of size 19, values of r_s numerically as large as 0.388 may be expected by chance at the 5% level of significance using a one-tailed test. The value we have obtained is much less than this and hence again there is no reason from this data to reject the hypothesis of zero correlation.

Exercises

14.1   A violinist was asked to place 10 given notes in order of increasing pitch. The correct order was: 2   6   4   5   3   7   9   1   8   10

The order of the violinist was: 3   5   4   6   2   8 10   1   7   9
Is it likely that such a measure of agreement could have been obtained by chance?

14.2 At a certain university a professor was responsible for giving a series of lectures to a large number of students, setting the examination and marking the papers, but otherwise he had no contact with students. Tutors were responsible for personal coaching of students in groups of 15 students per tutor. One tutor was selected at random and asked to rank his 15 students in decreasing order of ability in the subject. After the final examination these rankings were compared with the examination performance with the following result:

Examination   8 12   4   6 11 15 10   7   3   9   2 14   1 13   5
By tutor   5 15   7   4   8 12 10 11   1   9   3 13   2 14   6

Does this suggest that examinations are unnecessary?

14.3 Explain how the methods of this chapter could be used to rank the tutors in Exercise *14.2* in order of their ability to forecast examination results.

14.4 With manpower control in wartime, a certain number of the male staff of each firm was regarded as being essential to industry and was released from call-up. The men were accordingly ranked in decreasing order of their 'indispensability' to the firm. These rankings were later compared with their position on the salary scale, with the following results:

Name	Indispensability Ranking	Salary per annum ($)
James	1	10,000
Smith	2	7,500
Brown	3	7,500
Thomas	4	5,000
Hans	5	4,000
Ellis	6	4,750
Flower	7	5,000
Banks	8	6,000
Cohen	9	5,500
McDonald	10	5,250
Mitchell	11	4,600
Edwards	12	5,000
Drane	13	4,600
Barton	14	6,000
Oliver	15	5,400

Would you describe the salary scale as a 'fair' one in view of the above?

15

Estimation

One purpose of taking a sample is to use the statistics from it to estimate the parameters of the population, that is, the characteristics which describe the population distribution. We have mentioned earlier that we can use $\bar{x}$ from a sample as an estimate of the mean of the population. We can use the proportion p of successes in a sample to estimate the proportion of successes in the population distribution. Generally we would expect that the larger the sample, the better the estimate. Accordingly when we make an estimate we would normally ask ourselves three questions.

1. How accurate is it?

2. How big a sample would we require to produce a given accuracy?

3. Within what range might we confidently expect the population parameter to lie?

Let us consider some problems in which we attempt to answer these questions.

Example 15.1

Estimating the mean of a normal population (*given σ*)

An examiner who for many years has marked a state-wide final secondary school examination in mathematics has found from experience that the marks are approximately normally distributed. The mean varies from year to year, but the standard deviation remains remarkably constant at 12. He is just about to commence marking this year's papers and would like to know the mean mark that will result so that he can set limits for each grade of pass. To estimate the mean he marks 100 papers and finds $\bar{x} = 56$.

1. How accurate is 56 as an estimate of the population mean μ?

2. If he is not satisfied with this accuracy, how many more papers should he mark to be 95% confident that his estimate is not in error by more than 1 mark?

3. If he only marked the 100 papers within what range could he expect, with 95% confidence, the population mean to lie?

Solution

1. We know that the mean $\bar{x}$ is normally distributed with mean μ and standard deviation $= \dfrac{12}{\sqrt{100}} = 1.2$ marks. Hence 95% of samples will have a mean $\bar{x}$ within 1.96 standard deviations (i.e. within 2.4 marks) from μ. That is, only one sample in twenty will have a mean which differs from μ by more than 2.4.

This particular sample could be the one sample in twenty which differs by more than 2.4, and so all that we can say is that we are 95% confident that the population mean lies within 2.4 marks from 56.

2. Let n be the total number required in the sample.

Then the standard deviation of the mean $= \dfrac{12}{\sqrt{n}}$

Therefore 95% of samples will have a mean within $\dfrac{1.96 \times 12}{\sqrt{n}}$ from μ.

Hence he requires

$$\frac{1.96 \times 12}{\sqrt{n}} = 1$$

$$\text{giving} \quad n = 553$$

He therefore needs to mark a further 453 papers to be 95% confident that his estimate is not in error by more than 1 mark.

3. In *1.* it was pointed out that we are 95% confident that the population mean lies within 2.4 marks of 56. That is, we are 95% confident that μ lies within the range 53.6 to 58.4. We call this the 95% confidence interval for μ and the end points 53.6 and 58.4 are called 95% confidence limits for μ.

Example 15.2

Estimating the mean of a population (when (i) σ is NOT given and (ii) the sample is LARGE)

If in the previous example a complete new syllabus and a completely new type of examination had been adopted, the examiner could not lean on past experience as a guide, i.e. he would not know how marks were likely to be distributed nor would he know the standard deviation. He would have to base his estimates purely on information derived from marking the 100 papers. If the 100 marks had a mean of 56 and a standard deviation of 10 what would be the 95% confidence interval for the population mean mark based on this sample?

Solution

Although we do not know how the marks are distributed, because of the Central Limit Theorem we may assume $\bar{x}$ to be normally distributed with mean μ in the case of a sample of this size. Also, as we pointed out on page 101, we may use $\dfrac{s}{\sqrt{n}}$ in the case of large samples as an estimate of the standard deviation of $\bar{x}$. Hence this standard deviation here equals $\dfrac{10}{\sqrt{100}}$ or 1.

Ninety-five percent of samples will therefore have a mean which lies within 1.96×1 or 1.96 marks of μ. We have one sample whose mean is 56. Hence the 95% confidence interval for μ based on this sample is from 54.04 to 57.96, say 54.0 to 58.0.

Example 15.3

Estimating the mean of a NORMAL population (when (i) σ is NOT given, and (ii) the sample is SMALL)

In the case of small samples we are not justified in estimating the standard deviation of $\bar{x}$ from the variation within the sample itself. We can however use the t distribution to obtain confidence limits *but only if we can reasonably assume that the population is normally distributed.*

If in *Example 15.2* only 15 papers had been marked, and if we were satisfied from past experience that marks are normally distributed, we could use the fact that $\dfrac{\bar{x}-\mu}{s}\sqrt{n-1}$ follows a t distribution to obtain confidence limits. Given that for this sample $n = 15$, $\bar{x} = 56$ and $s = 10$ what are the 95% confidence limits for μ?

Solution
From *t* tables we find that 95% of samples, with 14 degrees of freedom, have a value of *t* numerically less than 2.14.

Hence for 95% of samples $\dfrac{\bar{x} - \mu}{10} \sqrt{14}$ or $.3742(\bar{x} - \mu)$ is numerically less than 2.14. That is, for 95% of samples $\bar{x} - \mu$ is numerically less than 5.72. Hence from our sample which has $\bar{x} = 56$, we obtain the 95% confidence limits for μ as 50.27 and 61.73 or roughly $50\tfrac{1}{4}$ and $61\tfrac{3}{4}$.

This method can, of course, be used for samples of any size, but only if the population is normally distributed. For large samples this method is in effect the same as that in *Example 15.2* because the *t* distribution approaches the normal as *n* becomes large.

Example 15.4

Estimating a population proportion
A random sample of 400 individuals yielded 49 who had completed some form of tertiary education.

1. How accurate is this sample proportion $\dfrac{49}{400}$ as an estimate of the proportion π who have completed tertiary education in the population?
2. How large a sample would be needed to be 95% confident that the error of the estimate would not exceed 0.02?
3. What are the 95% confidence limits for π?

Solution
1. This is a binomial situation since individuals either have, or have not, completed some form of tertiary education. Therefore the number in samples of 400 who have completed such education would be approximately normally distributed with mean 400π and standard deviation $\sqrt{400\pi(1-\pi)}$. The *proportion* who have completed tertiary education in samples of 400 is $\dfrac{1}{400}$ times the actual number. This proportion is therefore approximately normally distributed with mean π and standard deviation $\dfrac{1}{400}\sqrt{400\pi(1-\pi)}$ or $\sqrt{\dfrac{\pi(1-\pi)}{400}}$.

Hence in 19 samples out of 20 the proportion in the sample will differ from the population proportion π by less than $1.96\sqrt{\dfrac{\pi(1-\pi)}{400}}$.

Ours may be the one unusual sample in 20, but we are 95% confident that the error in using $\dfrac{49}{400}$ as an estimate of π is less than $1.96\sqrt{\dfrac{\pi(1-\pi)}{400}}$. However we do not know the value of π to substitute in this formula. With large samples we can find the error approximately by using for π the proportion in the sample. This gives

$$1.96\sqrt{\frac{49}{400}\times\frac{351}{400}\times\frac{1}{400}}=.033$$

Hence we are 95% confident that π will not differ from the sample value of 0.1225 by more than .033.

2. In this case we require a sample of size n such that

$$1.96\sqrt{\frac{\pi(1-\pi)}{n}}=0.02.$$

Using the sample value of $\dfrac{49}{400}$ as an approximation for π in this formula we obtain by squaring both sides

$$n=\frac{1.96\times1.96\times\dfrac{49}{400}\times\dfrac{351}{400}}{.02\times.02}=1032$$

A sample of size 1032 is needed.

3. It was pointed out in *1.* that we are 95% confident that the population proportion lies within .033 of 0.1225. Hence the 95% confidence limits for π are 0.0895 and 0.1555 or say .089 and .156.

Confidence intervals for other parameters

These methods may be extended to determining confidence intervals for the estimate of the difference between two means or the difference between two proportions. More complicated methods are available for determining the confidence intervals for σ and other parameters but these will not be discussed here.

Exercises

15.1 A school physical training teacher wished to determine the mean height which 7 year old students can jump. In view of the length of time which such an experiment would take with many students, he decided to select a random sample of 12. The heights jumped (in centimetres) were

$$44 \quad 52 \quad 40 \quad 42 \quad 46 \quad 53 \quad 43 \quad 50 \quad 60 \quad 56 \quad 54 \quad 48$$

a. Estimate the population mean.
b. How accurate is this estimate?
c. Find the 95% confidence limits for the population mean.

15.2 If the teacher in *15.1* had had time to test a sample of 100 students and had obtained from the sample a mean height of 48 cm with a standard deviation of 5 cm., what would then be the answers to *a.*, *b.* and *c.* of Exercise *15.1*?

15.3 Out of a random sample of 400 people who were asked to state either whether they were in favour, or not in favour, of a certain contentious television programme, 256 people were recorded as 'in favour'.
a. What are the 95% confidence limits for the proportion in favour in the population?
b. How large a sample would be required to be 95% confident that the error in the estimate does not exceed 0.02?

Appendix I

1 :   **Miscellaneous Problems**

2 :   **Solutions to Exercises**

3 :   **Answers to Miscellaneous Problems**

1: Miscellaneous Problems

Note: (*1*) *The problems are not in any special order.*

(*2*) *Some of the problems are quite difficult.*

1. A company selling rope finds that during the day the following lengths of rope were sold (measured in metres):

$$4, 6, 5, 3, 9, 8, 9, 4, 6, 6.$$

a. Calculate the mean, the median and the standard deviation of these lengths.

b. If these lengths had been measured in centimetres they would, of course, have been

$$400, 600, 500, 300, 900, 800, 900, 400, 600, 600.$$

What is the mean, the median and the standard deviation of the lengths measured in centimetres?

2. Heights of pine trees in a certain forest are normally distributed with a mean of 48 metres and a standard deviation of $2\frac{1}{2}$ metres. What is the probability that a tree chosen at random should have a height

a. Above 53 metres

b. Between $45\frac{1}{2}$ metres and 49 metres

c. Below 44 metres?

d. If you inspected 200 of the trees, how many of them would you expect to be over 50 metres tall?

3. Paper clips produced in a certain factory have a length which is approximately normally distributed with a mean of 2.93 cm, and a standard deviation of 0.02 cm. In a batch of 5000 clips, how many would you expect to be

a. Longer than 2.96 cm

b. Between 2.92 and 2.946 cm ?

4. If, in the batch of 5000 paper clips referred to in problem *3* it was found that 1500 were longer than 2.96 cm, could you draw any conclusion?

5. The marks obtained in a class examination were

5 4 5 7 6 8 3 6 8 6 4 5 7 5 6 9 6 7 5 8

a. Set the marks out in the form of a frequency table
b. Calculate their mean and standard deviation from the frequency table.

6. Suppose that a coin is tossed 10,000 times. Determine an upper bound and a lower bound for the number of heads, so that the probability is 0.95 that the number of heads which does appear lies between these bounds.

7. The probability of catching your train in the morning is 0.5, and of catching your train in the evening is 0.9. Assume these are independent events. What are the probabilities of

a. Missing the train both in the morning and the evening;
b. Catching it both times
c. Missing it at most once
d. Catching it at least once?

8. The scores obtained by individuals in a certain experiment have been found to be normally distributed, with mean 60 and standard deviation 10.
a. The scores obtained by four individuals were 58, 70, 90 and 52. Express these performances in the form of standard scores.
b. What is the probability of obtaining by chance from this population an individual whose score
  (*i*) Is 67 or greater
 (*ii*) Lies between 50 and 64
(*iii*) Is 45 or less
 (*iv*) Is either 52 or less, or 72 or more?
c. Students are divided into groups of 4 for this experiment and the mean score for each group is determined. How would these group means be distributed?
d. Are we justified in considering the performance of the group in *a.* as unusual?

9.*a.* A manufacturer has been asked to produce a number of cylindrical objects which will pass through a hole 1 cm in diameter. The objects produced have a diameter which is approximately normally distributed with mean 0.85 cm and standard deviation 0.1 cm.

Given a randomly chosen object
(*i*) What is the probability that it will fit through the hole?
(*ii*) What is the probability that the diameter of the hole will exceed that of this object by more than 0.12 cm ?
b. Given two such objects, what is the probability that
(*i*) They will both fit through the hole
(*ii*) At least one will fit through the hole?

10. The length of time taken to service a particular car is normally distributed with mean 5 hours and standard deviation 1 hour. If the car is left at the service station at 10 a.m., what is the probability that
a. It will be finished before 5 p.m.
b. It will not be finished by 5.30 p.m., but will be ready by 6 p.m.?

11. Two different lecturers decided to estimate the likely results of the students in a class of ten students. Their assessments (expressed as a mark out of 100) are set out in the following table:

Student	Lecturer	
	Green	Brown
Smith	84	73
Jones	80	84
Watson	74	80
Holmes	69	71
Taylor	63	70
Burton	59	60
Harvey	55	65
O'Neill	52	50
Prince	41	54
King	25	19

a. Are these assessments consistent with the assumption that neither lecturer has a real tendency to give students generally a higher assessment than his colleague?
b. Is the superficial similarity of ranking in order of merit by the two lecturers likely to have happened without any real agreement between them?

12. Of 100 patients admitted to a hospital and treated by method A, 20 died. Of 400 patients admitted to the same hospital and treated by method B, 40 died. Is the difference likely to have happened by chance?

13.a. In a certain area, 75% of males born attain age 60. In a randomly selected group of 1000 male babies, what is the probability that no more than 710 will attain age 60?

b. In 1965, a random sample of 10,000 names was taken from the register of births for the year 1905, and it was found after much investigation that only 7,345 of these people were still alive. Does it appear that the death rate for this group of people has been higher than might reasonably have been expected on the assumption that 75% would live to age 60?

14. A regular social tennis player has noted that the average length of a rally (the length being defined as the number of times the ball crosses the net in the course of one point) is 9, with a standard deviation of 2.
a. What, then, is the probability that the average length of 25 rallies should be greater than 10?
b. This player went to watch the professionals play, and watched 484 rallies whose mean length was only 7. Comment.

15. The number of left-handed and right-handed teenagers in a certain community who were being treated for stuttering and the numbers who were not being treated are set out in the following table:

	Being treated	Not being treated	Total
Left-handed	18	182	200
Right-handed	82	1,718	1,800
Total	100	1,900	2,000

Can we conclude that left-handers are more liable to be stutterers?

16. The following is an actual record of the scores in 15 scrabble games between the same two players:

Player 1	203	227	225	202	235	256	257	289
	245	224	318	325	288	278	203	
Player 2	215	236	244	231	165	256	169	214
	205	259	232	192	215	261	215	

a. Is there any conclusive evidence of any real difference between these two players?
b. It is stated in the instructions for playing scrabble that most games between reasonably competent players result in a total point score of (about) 500. Is there evidence from the above scores that these two

players have a lower average total score than this (and are therefore either less than 'reasonably' competent, or else they are 'defensive' players)?

17. A school divides its final year students in mathematics into three classes of 10. The students' marks in the final examination are shown in the table below:

Class	Marks									
A	23	37	36	39	33	22	31	37	23	40
B	21	28	30	29	34	33	20	27	39	38
C	17	39	30	23	31	26	18	27	20	18

a. What is the median mark for the school?
b. Are the three classes likely to be random samples from the same population with this median?

18. For the marks in problem *17.*
a. Draw a less than ogive
b. Determine from the ogive the median, the upper and lower quartiles and the ninety percentile.

19. A cat fancier has estimated that the chances of a cat having a litter of a particular size are as follows:

Size of litter	Probability
1	1/20
2	1/10
3	1/5
4	2/5
5	3/20
6	1/20
7 or more	1/20

Over a year, she observed large numbers of cats, with the following results:

Size of litter	Number of litters
1	35
2	90
3	200
4	380
5	225
6	40
7 or more	30

Comment.

20.a. A certain lecturer in statistics often travels between Sydney and Orange by road, and being statistically minded, records his travelling time on each occasion. An examination of these records shows that this time is approximately normally distributed with mean 4 hours 5 minutes and standard deviation 16 minutes. On a particular day, he requires to be in Orange not later than 6.50 a.m. What is the latest time at which he can leave Sydney in order to be 95% sure that he will not be late?

b. Another regular traveller on this section has travelled over it 121 times, with a mean travelling time of 4 hours 22 minutes. Can we reasonably assert from this that he is a slower driver than the lecturer?

21. Experience has shown that steam locomotive boilers have an effective mean working life of 20 years, with a standard deviation of 3.5 years. A certain coal company on the northern coalfields runs its own private railway, and the nine engines in service have boilers an average of 22 years old. Comment.

22. A survey of the linguistic habits of school pupils has produced the following numbers using a short vowel and a long vowel in a particular word:

Numbers using a long vowel

	Private school		State School	
	Males	Females	Males	Females
City schools	156	134	215	228
Country schools	102	88	224	233

Numbers using a short vowel

	Private school		State School	
	Males	Females	Males	Females
City schools	207	213	192	231
Country schools	163	137	226	251

Write down the fraction of
a. Students that are males
b. Male students that attend city schools
c. Female students that use a long vowel
d. Male private school students that attend city schools

e. Students that attend country schools

f. Students that use a short vowel

g. Students that attend private schools

h. Students in state schools in the city that are females

i. Female students at city schools who use a long vowel

j. Male students at private schools in the city who use a short vowel.

23. A certain university course has both theoretical and practical aspects which can be tested separately. Ten students were ranked in the two sections as follows:

Rank in theory test	Rank in practical test
3	1
6	5
4	2
7	10
10	7
1	3
5	8
2	4
9	9
8	6

Is there any significant difference in the performance of the students in the two sections of the course?

24. It is known from the mathematical theory of statistics that if s is the sample standard deviation observed in a sample of size n from a normal population whose standard deviation is σ, then the quantity $\dfrac{ns^2}{\sigma^2}$ is distributed as χ^2 with $n - 1$ degrees of freedom. Use this fact to determine if an observed standard deviation of 12 from a sample of size 100 is consistent with the hypothesis that the population standard deviation is 10.

25. An employee due at work at 8.55 a.m. has been accused by his employer of being late too frequently. The employee contends that his routine enables him to get to work five minutes early on the average, and it is only random fluctuations which occasionally cause him to be late. In support of this, he submits the following table as a record of his arrival times for the last two weeks.

	Mon	Tues	Wed	Thur	Fri
Week 1	8.49	8.55	8.55	8.57	8.48
Week 2	8.44	8.53	8.50	8.50	8.57

Are these figures consistent with the employee's contention?

26. Transistors of a certain type are advertised to have a failure rate of 1 % during their first year of life. 1225 of these transistors were installed in a computer, and during the first year of operation 27 of them had to be replaced. Could the owner of the computer reasonably complain that the advertised failure rate was false?

27.a. A large group of people was asked to record the average length of their telephone conversations, and it was found that the mean length was 28 minutes 30 seconds, with a standard deviation of 4 minutes. However 25 female university students reported that their respective average lengths of call were (in minutes) 31, 47, 27, 19, 32, 33, 36, 33, 41, 18, 19, 25, 33, 38, 52, 12, 17, 23, 34, 35, 26, 27, 28, 14 and 50. Comment.

b. Would your comment have been different if you had *not* known that the standard deviation of the large group had been found to be 4 minutes?

28. On the average, 1 motorist in 20 gets booked for speeding when passing through a certain small country town. On a certain day, however, 32 of the 400 motorists who passed through the town were booked. Does it appear that the local policeman was unusually active on that day?

29.a. Haydn's Symphony No. 94, *The Surprise*, has been performed many times over the past 200 years or so, and it has been discovered that concert performances have lasted on the average 22 minutes 30 seconds, with a variation (due to conductors' preferences, audience reaction and other factors) which has produced a standard deviation of 1 minute. On a certain night, it was noticed that a performance of this symphony lasted 24 minutes. What comment can you make?

b. The critic who made this observation thought that this apparent slowness was due to the conductor's generally slow approach, so he went to 8 other performances of the same symphony with the same conductor, and found that the average time taken (over the nine performances in all) was 23 minutes 30 seconds. Does this reinforce or detract from the critic's belief that this conductor regularly produces a slower than normal performance?

30.a. It has been discovered that a certain train which is scheduled to leave a station at 7.44 a.m., in fact leaves there on the average at 7.46 a.m., with a standard deviation of 75 seconds. How often is the train actually early?

b. If a passenger wants to be 99% sure that he will catch the train, what is the latest time at which he should get to the station?
You may assume that the time of leaving is normally distributed.

31. Experience in teaching statistics has shown that students with a mathematical background take about three days, with standard deviation 1 day, to grasp the idea of a normal distribution. Most of the students doing an introductory statistics course have little mathematical background, and it would appear that a group of 50 students has taken a mean time of 4 days to grasp the idea. Can we reasonably conclude that students with an inadequate mathematical training are much slower than those who have some considerable mathematical experience?

32. A lecturer travels regularly between Macquarie University and Hurstville. For a while he travelled via The Boulevarde at Strathfield, and found that 90 times out of 120 it was necessary to stop at the traffic lights at the intersection of Hume Highway. He was unhappy about this, and now travels via Homebush Road. He has been delayed at the lights at the intersection of this road with Hume Highway on 49 occasions having made 98 trips over this route. Can he reasonably conclude that Homebush Road is a more satisfactory route than The Boulevarde, assuming that conditions on the two roads are equivalent except for the traffic lights?

33. The following are the scores per move made by one player in one of the scrabble games in problem *16*:

22 13 10 16 10 14 24 12 7 13 10 30 16 15 8 20 27 11

Are these results consistent with the assumption that the mean score per move for the player is 15?

34. The following table gives the results of a survey conducted in three universities to determine what proportion of students travel by car to the university:

	University A		University B		University C	
	Day	Evening	Day	Evening	Day	Evening
Regularly by car	30	150	40	90	25	70
Occasionally by car	110	110	100	170	65	120
Never by car	310	40	260	40	210	110
Totals	450	300	400	300	300	300

Test to see whether there is
a. Any significant difference in car travelling habits amongst the different universities, considering day and evening students separately
b. A significant difference between day and evening students within universities.

35. Prepare a diagrammatic presentation bringing out the salient features of the data in problem *34*.

36. The work of a certain student has been observed by taking a random sample of a small number of pages from his exercise book. One week, 20 pages were inspected, and an error was found on 17 of these. Next week, however, only 15 out of 30 pages contained an error. Can we conclude that the student has improved?

37. During one week, a tutor who should have a total of 92 students in his tutorial groups found that 80 of these were present, whereas another tutor who nominally has 95 students found that only 60 were present. Does it appear the first tutor is more popular with his students than the second?

38. As at June 1966, according to the *Monthly Review of Business Statistics*, the population of Victoria was 3,217,832, whereas that of NSW was 4,231,103. At this time, 7,839 people were receiving unemployment benefits in NSW, while the corresponding figure in Victoria was 3,450. Does it appear that the proportion of people receiving unemployment benefits in the two states is different?

39.a. Reports submitted to a certain organisation have a median length of 1,500 words, the lower quartile being 1,250 words. What is the probability that
 (*i*) Each of 10 reports should be longer than 1500 words
 (*ii*) At least one of 10 reports should be longer than 1500 words
 (*iii*) A report should be at least 1250 words, but not more than 1500
 (*iv*) Of two reports, one should be longer than 1500 words and the

other at least 1250, but not more than 1500?

b. If it is also known that the length of the reports is normally distributed, with the lower quartile and median as specified above, what is the standard deviation?

40. At a certain university, enrolments in certain faculties in a particular year were as set out in the following table:

	Arts	Science	Economics
Male	1830	1450	445
Female	2170	1150	255

Is there any interrelation between the different faculties and the sex ratios within the faculties?

41. It is known that a certain brand of 'superball' will bounce to a height of about 144 cm. if dropped to a hard surface from a height of 100 cm. A schoolboy discovered that if he immersed a superball in ethylene glycol, it appeared to bounce higher. He therefore bought ten balls, and before doing anything to them, bounced them from a height of 100 cm. and noted the heights to which they rose. Heights measured in centimetres were

144   136   148   152   150   160   148   146   158   142

He then treated them with ethylene glycol and repeated the bounces. The heights then observed were (recorded in the same order as before)

148   140   160   164   148   164   156   160   164   152

Do you think it is worth the trouble for him to patent ethylene-glycol-treated superballs?

42. A remote automatic lighthouse has two lanterns, burning together. The life of a lantern is approximately normally distributed with mean 800 hours and standard deviation 50 hours.

a. If the lanterns are replaced every 4 weeks (672 hours) what is the probability that both lanterns will still be burning when replaced?

b. If it is decided that satisfactory performance is obtained if at least one lantern is burning, how often should the lamps be replaced so that unsatisfactory performance will occur no more than once in every 100 replacements on the average?

43. Calculate the mean, standard deviation and mean deviation for the following data, representing the number of correct answers obtained by students in a certain examination.

Number of correct answers	Number of students
1	41
2	44
3	48
4	61
5	74
6	60
7	44
8	49
9	53
10	26

44. The following table sets out the number of students in a university cafeteria at a specified time over a three-week period.

Day	Number of males	Number of females
M	54	56
T	55	58
W	57	60
T	62	54
F	63	55
M	57	58
T	64	65
W	60	59
T	61	60
F	62	60
M	63	58
T	56	59
W	60	58
T	62	56
F	64	54

a. Calculate the product-moment correlation coefficient. (*Note:* If a suitable arbitrary origin is chosen the arithmetic only involves small numbers and the calculations are easy.)

b. Why is this coefficient not appropriate here for testing whether the correlation in the population is zero?

c. Calculate the rank correlation coefficient and use the value obtained to test the hypothesis that the correlation in the population is zero.

45. In a manufacturing process which has been in operation for many years, the lengths of one of the items produced have been found to follow an approximately normal distribution with a mean of 45.00 mm and a standard deviation of 0.25 mm. Recently, however, an on-line computer has been installed to assist in process control, and this (together with replacement of some components in the original machinery) is thought to have affected the output. To test this, a sample of 65 items was taken and this was found to have a mean of 44.87 mm and a standard deviation of 0.21 mm. Do you consider that the changes in the production scheme have really resulted in any change in the characteristics of the output? *Hint:* The methods in this book can be used to test the mean. You can test the standard deviation by making use of the theory given in problem *24.*

46.a. At the time that this problem is being constructed, a Roller Game series is in progress between the Australian Thunderbirds and the San Francisco Shamrocks, and the T'birds have won all fourteen games so far played. If the teams are really evenly matched, what is the chance that such a result should occur by chance alone?

   b. The following gives a detailed record of four games between these teams: (Each game consists of eight twelve-minute periods, periods 1, 3, 5 and 7 being skated by the girls and 2, 4, 6 and 8 by the men.)

Game No.	Team	Periods								Total
		1	2	3	4	5	6	7	8	
1	T'birds	10	6	15	15	0	7	13	11	77
	S'rocks	10	8	9	19	8	5	0	15	74
2	T'birds	6	12	13	11	16	6	8	31	103
	S'rocks	5	5	5	15	6	34	16	15	101
3	T'birds	10	17	10	9	8	15	10	21	100
	S'rocks	9	11	8	16	9	18	8	16	95
4	T'birds	10	11	10	5	20	20	9	35	120
	S'rocks	9	12	10	36	10	14	10	9	110

(i) Do these detailed figures confirm that the T'birds are superior to the Shamrocks?

(ii) Is there a tendency for the girls' teams to score fewer points than the men?

(iii) Is there a significant correlation between the two teams' scores?

(iv) A superficial inspection suggests that there is a tendency for the first period of each game to contain relatively low scores and the final

period of each game relatively high scores. Is this impression justified?
(*v*) Is there any tendency for more points to be scored in the second half of a game than in the first half?
(*vi*) Further to (*i*), can you say anything about the superiority of the T'birds girls relative to the Shamrock girls, and of the T'bird men relative to the Shamrock men?
(*vii*) If one team is ahead in points already scored up to a certain point in the game, is there a tendency for the other team to score the greater number of points in the next period?
Hint:Calculate the progress scores in each game after each period, and note which team is behind. Note a + if that team scores more than the other team in the next period. Is the proportion of + s excessively high?

1.1a. $3A - 4B = (3)(2) - (4)(-3)$
$$= 6 + 12$$
$$= 18$$

b. $6A + 5B + 2 - 4D = (6)(2) + 5(-3) + 2 - (4)(-2)$
$$= 12 - 15 + 2 + 8$$
$$= 7$$

c. $\dfrac{AB + CD}{AC - BD} = \dfrac{(2)(-3) + (4)(-2)}{(2)(4) - (-3)(-2)}$
$$= \frac{-6 - 8}{8 - 6}$$
$$= \frac{-14}{2} = -7$$

d. $A^2 + B^2 - 4CD = 2^2 + (-3)^2 - (4)(4)(-2)$
$$= 4 + 9 + 32$$
$$= 45$$

e. $3(A + 2C) - 2(B - 4D) = 3[2 + (2)(4)] - 2[-3 - (4)(-2)]$
$$= (3)(10) - (2)(5)$$
$$= 30 - 10 = 20$$

$f.$

$$\frac{B^2 - C^2}{BC - 1} = \frac{(-3)^2 - (4)^2}{(-3)(4) - 1}$$

$$= \frac{9 - 16}{-12 - 1}$$

$$= \frac{-7}{-13} = \frac{7}{13}$$

$g.$

$$\sqrt{2A^2 - B^2 + C^2} = \sqrt{2(2)^2 - (-3)^2 + (4)^2}$$

$$= \sqrt{8 - 9 + 16}$$

$$= \sqrt{15} = 3.873 \text{ from square root tables.}$$

$h.$

$$\sqrt{\frac{2A^2}{B^2} + \frac{3C^2}{D^2}} = \sqrt{\frac{2 \times (2)^2}{(-3)^2} + \frac{3(4)^2}{(-2)^2}}$$

$$= \sqrt{\frac{8}{9} + \frac{48}{4}}$$

$$= \sqrt{12.89} = 3.59$$

1.2 We are given $Y = X^2 - 2X - 5.$

Therefore when $X = -3$ by substitution $Y = 10$

$X = -2$ by substitution $Y = 3$

$X = -1$ by substitution $Y = -2$

$X = 0$ by substitution $Y = -5$

$X = 1$ by substitution $Y = -6$

$X = 2$ by substitution $Y = -5$

$X = 3$ by substitution $Y = -2$

$X = 4$ by substitution $Y = 3$

$X = 5$ by substitution $Y = 10$

Hence the graph is

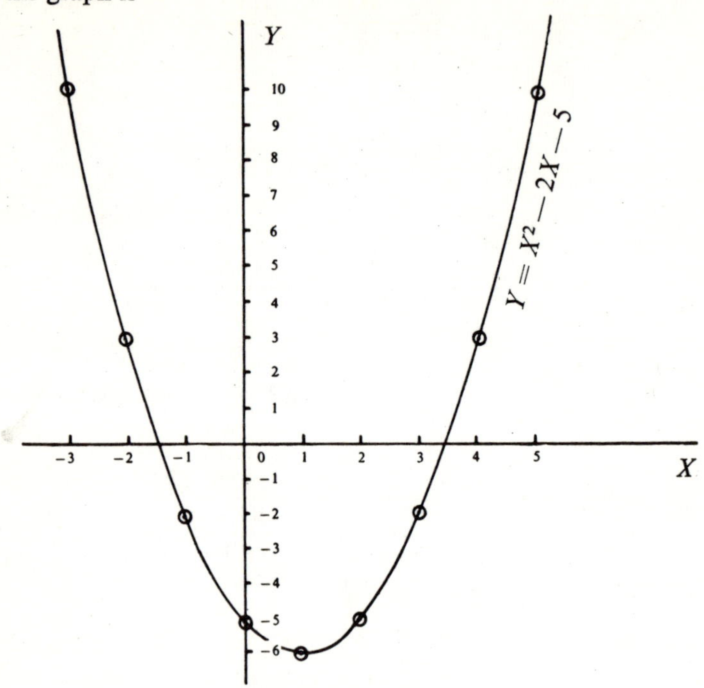

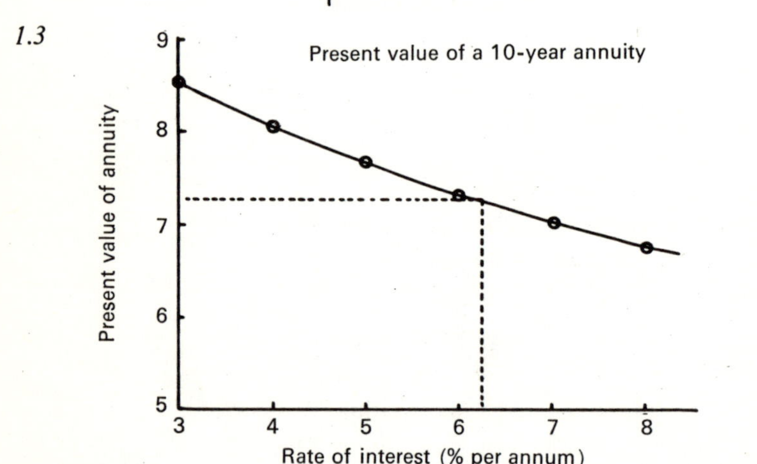

1.3

Estimated present value $= 7.27$

*Note:*This is not the most accurate way of estimating the value.

1.4a.

$$\sqrt{\frac{(M-P)^2}{R} + \frac{(L-M)^2}{N}} = \sqrt{\frac{(3+\frac{1}{3})^2}{\frac{1}{6}} + \frac{(-2-3)^2}{\frac{1}{2}}}$$

$$= \sqrt{\frac{(100)(6)}{9} + (25)(2)}$$

$$= \sqrt{116.667}$$

$$= 10.80 \text{ from square root tables.}$$

b.

$$\frac{R-P}{\sqrt{L^2+M^2}}(L^2R[N+P]) = \frac{\frac{1}{6}+\frac{1}{3}}{\sqrt{(-2)^2+3^2}}(-2)^2(\tfrac{1}{6})[\tfrac{1}{2}-\tfrac{1}{3}]$$

$$= \frac{1}{2} \cdot \frac{1}{\sqrt{13}} \cdot \frac{4}{6} \cdot \frac{1}{6}$$

$$= 0.0154$$

1.5a.

$$x = \frac{(0.05632)(579.2)}{(12.32)(0.003875)}$$

$$\log x = \log 0.05632 + \log 579.2 - \log 12.32 - \log 0.003875$$

$$= -2 + .7507 + 2 + .7628 - 1 - .0906 - (-3)$$
$$-.5883$$

$$= 2 + .8346$$

$$\therefore x = 683.3$$

b.

$$x = \sqrt{\frac{(46.84)(0.00752)^3}{(2.264)^5}}$$

$$\therefore \log x = \frac{1}{2}\log 46.84 + \frac{3}{2}\log 0.00752 - \frac{5}{2}\log 2.264$$

$$= \frac{1}{2}(1+.6706) + \frac{3}{2}(-3+.8762) - \frac{5}{2}(0+.3549)$$

$$= 0.8353 - 3.1857 - 0.8873$$

$$= -3.2377$$

$$= -4 + .7623$$

$$\therefore x = 0.0005785$$

$$c. \qquad x = \frac{39.42}{0.2364} \sqrt{\frac{(52.61)(0.07461)}{(0.02865)(0.6829)}}$$

$$\therefore \log x = \log 39.42 + \tfrac{1}{2} \log 52.61 + \tfrac{1}{2} \log 0.07461$$
$$- \log 0.2364 - \tfrac{1}{2} \log 0.02865 - \tfrac{1}{2} \log 0.6829$$
$$= 1 + .5957 + \tfrac{1}{2}(1 + .7211) + \tfrac{1}{2}(-2 + .8728)$$
$$- (-1 + .3736) - \tfrac{1}{2}(-2 + .4571) - \tfrac{1}{2}(-1 + .8344)$$
$$= 3 + .3733$$
$$\therefore \qquad x = 2362$$

$$1.6a. \quad \sum_{1}^{5} X_r = 2 + 4 + 3 - 2 + 6 = 13$$

$$b. \quad \sum_{1}^{5} Y_r = 5 - 3 + 1 - 3 + 4 = 4$$

$$c. \quad \sum_{1}^{5} X_r^2 = 4 + 16 + 9 + 4 + 36 = 69$$

$$d. \quad \left(\sum_{1}^{5} X_r\right)^2 = 13^2 = 169 \text{ from } a.$$

$$e. \quad \sum_{1}^{5} Y_r^2 = 25 + 9 + 1 + 9 + 16 = 60$$

$$f. \quad \left(\sum_{1}^{5} Y_r\right)^2 = 4^2 = 16 \text{ from } b.$$

$$g. \quad \sum_{1}^{5} X_r Y_r = 10 - 12 + 3 + 6 + 24 = 31$$

$$h. \quad \sum_{1}^{5} X_r^2 - \frac{1}{5}\left(\sum_{1}^{5} X_r\right)^2 = 69 - \frac{1}{5} \times 169 \text{ from } c \text{ and } d.$$
$$= 35.2$$

$$i. \quad \sum_{1}^{5} Y_r^2 - \frac{1}{5}\left(\sum_{1}^{5} Y_r\right)^2 = 60 - \frac{1}{5} \times 16 \text{ from } e \text{ and } f.$$
$$= 56.8$$

$$j. \quad \sum_{1}^{5} X_r Y_r - \frac{1}{5}\left(\sum_{1}^{5} X_r\right)\left(\sum_{1}^{5} Y_r\right) = 31 - \frac{1}{5} \times 13 \times 4 \text{ from } a, b \text{ and } g.$$
$$= 20.6$$

1.7

a.   *b.*

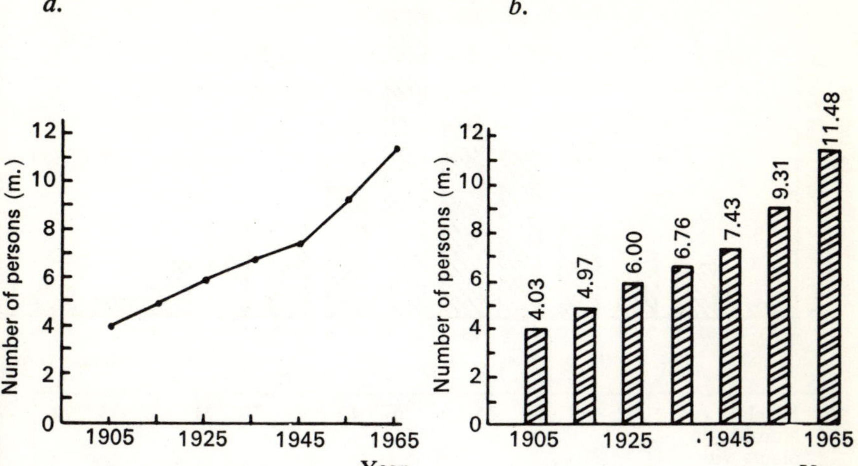

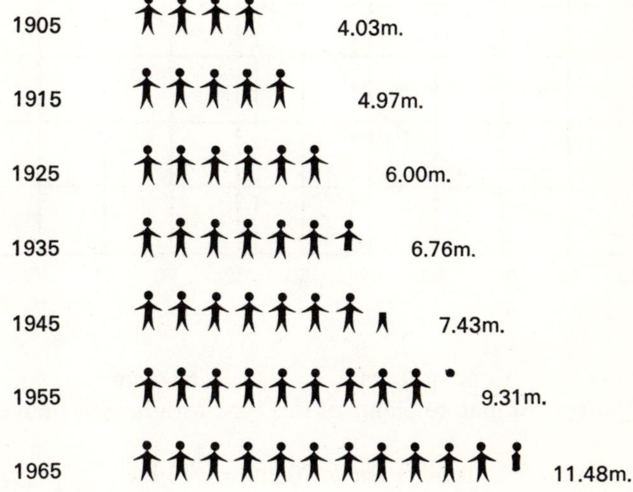

The population of Australia 1905-65 (each figure represents 1m. people).

1.8

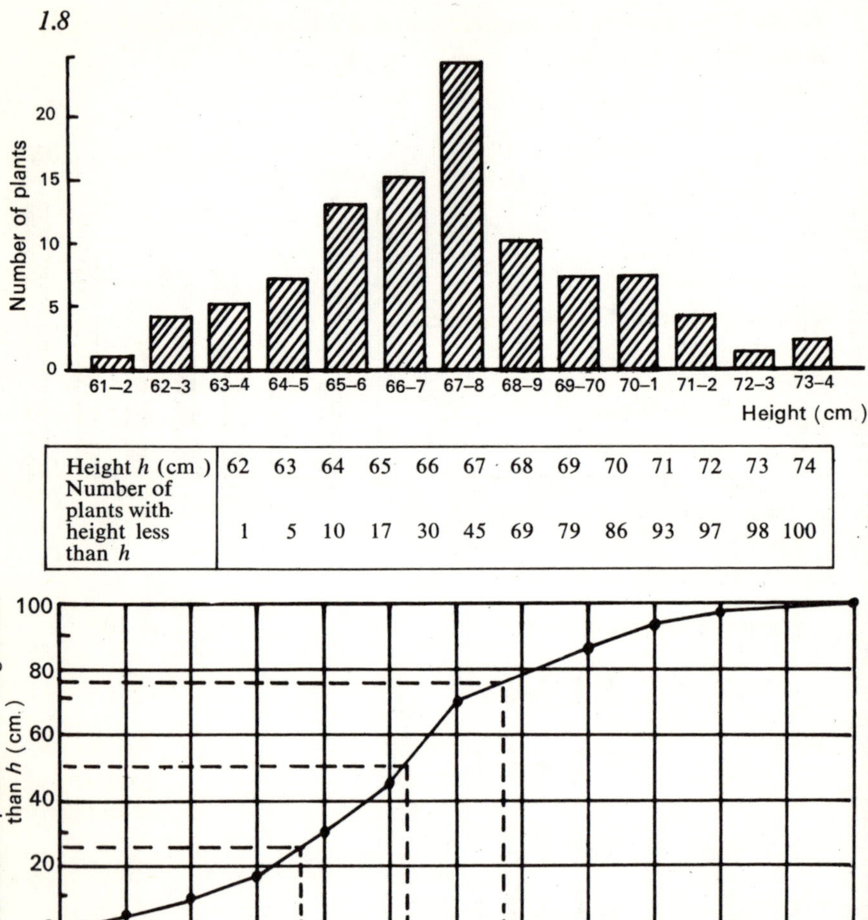

Height h (cm)	62	63	64	65	66	67	68	69	70	71	72	73	74
Number of plants with height less than h	1	5	10	17	30	45	69	79	86	93	97	98	100

a. $67\frac{1}{4}$ cm   *b.* 68.7 cm   *c.* 65.6 cm

The proportion of mature plants of this type with heights above 66 cm
= 0.7.

The proportion with heights above 70 cm = 0.14.

Hence the proportion with heights between 66 cm and 70 cm

$$= 0.7 - 0.14$$
$$= 0.56.$$

2.1 So many answers are required that it is quicker to arrange the data in the form of an array.

Thus:

46	46	47	48	48	49	49			
50	50	50	51	51	52	52			
53	53	53	54	54	55	56	56	56	56
57	57	58	58	58	58	59	59		
60	60	60	61	61	62	62	62		
63	63	63	63	64	65	65	65		
66	66	67	67	67	68	68	69		
70	70	71	72						

We can now write down the answers.

a. The greatest height $= 72$ cm

b. The lowest height $= 46$ cm

c. The range of heights $= 72 - 46$ $= 26$ cm

d. The eighth highest $= 67$ cm

e. The top four jumps were 70, 70, 71 and 72 cm

f. The number of jumps above 63 cm $= 16$

g. The number of jumps 53 cm or less $= 17$

h. The percentage of jumps above 54 cm

  but not above 64 cm $= \dfrac{26}{60} \times 100$

$= 43\tfrac{1}{3}\%$

2.2

Height jumped (cm)	Number of students	Relative frequency percentage
$45\tfrac{1}{2}$ — under $49\tfrac{1}{2}$	7	11.7
$49\tfrac{1}{2}$ — under $53\tfrac{1}{2}$	10	16.7
$53\tfrac{1}{2}$ — under $57\tfrac{1}{2}$	9	15.0
$57\tfrac{1}{2}$ — under $61\tfrac{1}{2}$	11	18.3
$61\tfrac{1}{2}$ — under $65\tfrac{1}{2}$	11	18.3
$65\tfrac{1}{2}$ — under $69\tfrac{1}{2}$	8	13.3
$69\tfrac{1}{2}$ — under $73\tfrac{1}{2}$	4	6.7
Total =	60	100.0

a. (*i*) 4 cm

(*ii*) $53\tfrac{1}{2}$ cm

(*iii*) 9

(*iv*) $\dfrac{9}{60} = 15\%$

(v) Both $57\frac{1}{2}$ — under $61\frac{1}{2}$, and $61\frac{1}{2}$ — under $65\frac{1}{2}$

(vi) $\dfrac{34}{60} \times 100 = 56\frac{2}{3}\%$

(vii) $\dfrac{22}{60} \times 100 = 36\frac{2}{3}\%$

b.(i)

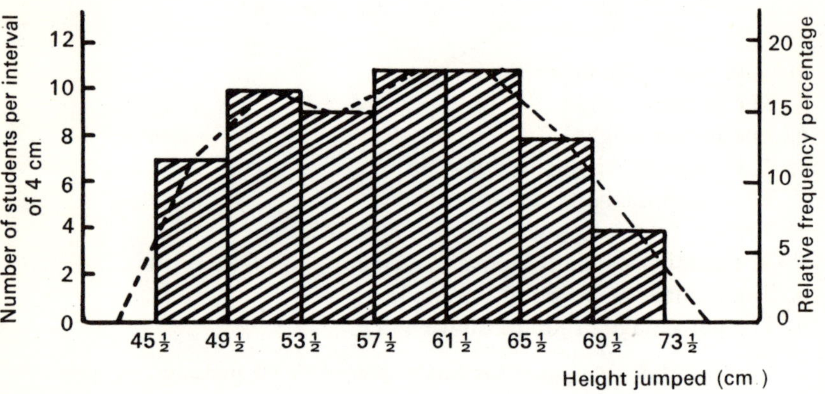

Histogram of heights jumped by 60 students.

(ii) and (iii). The relative frequency distribution table and the relative frequency polygon show the percentages in each class interval. The former is the third column which was added to the frequency table. For the latter we may use the histogram with the scale changed to that on the right hand side.

(iv) and (v).

Height jumped (cm)	Cumulative frequency	Cumulative frequency percentage
$45\frac{1}{2}$ — under $49\frac{1}{2}$	7	11.7
$49\frac{1}{2}$ — under $53\frac{1}{2}$	17	28.3
$53\frac{1}{2}$ — under $57\frac{1}{2}$	26	43.3
$57\frac{1}{2}$ — under $61\frac{1}{2}$	37	61.7
$61\frac{1}{2}$ — under $65\frac{1}{2}$	48	80.0
$65\frac{1}{2}$ — Both $69\frac{1}{2}$	56	93.3
$69\frac{1}{2}$ — under $73\frac{1}{2}$	60	100.0

(*vi*) and (*vii*).

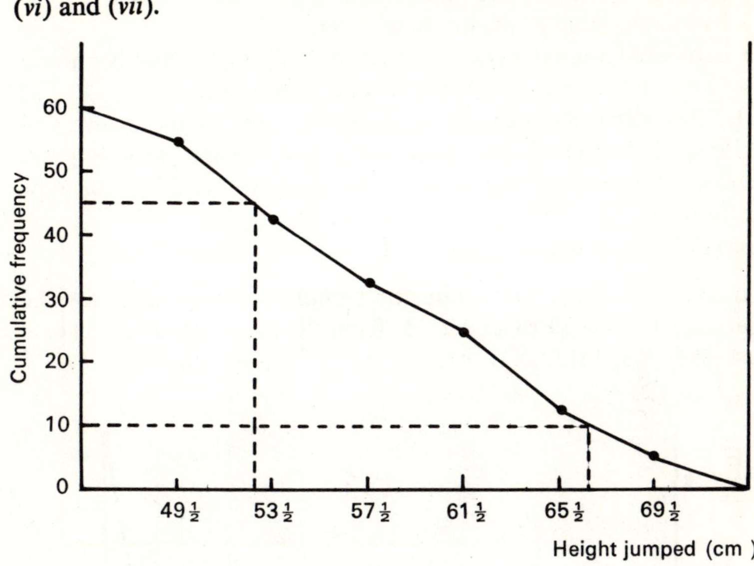

More than ogive for height jumped by 60 students.

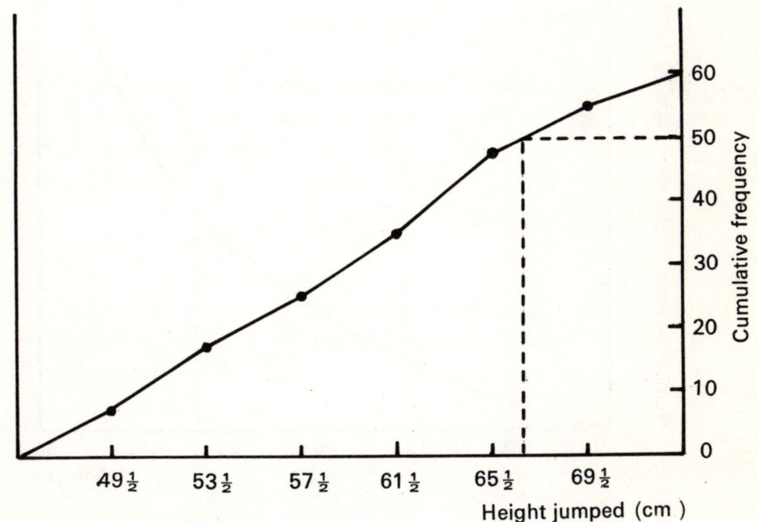

Less than ogive for height jumped by 60 students.

c. (*i*) Estimated number less than 67 cm   = 50
　(*ii*) Estimated number greater than 53 cm   = 45
　(*iii*) Estimated number between 53 cm and 67 cm = Number above
　　53 cm minus number above 67 cm = 45 — 10   = 35

Note In this particular exercise it is easy to obtain the answers to part *c.* from the original data. Frequently however we are only given the grouped data and we then need to use the ogive to answer this type of question.

Exercise 2.3 is such a case.

2.3 To draw the ogive we require the cumulative frequencies for the numbers aged from 30 to under 35, from 30 to under 40 etc. These are 164, 395, 785, 1413, 2421, 4133.

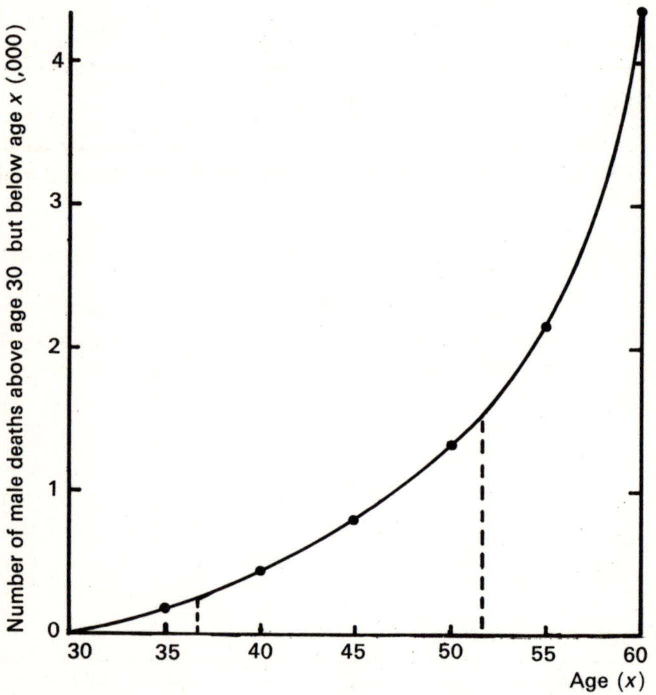

Less than ogive for male deaths in Australia in 1967.

The number over 30 but under age 37 estimated from the ogive is 250. The number over 30 but under 52 estimated from the ogive is 1770. Hence our estimate of the number of male deaths aged from 37 to under 52 is 1770 — 250 or 1520.

3.1a. Mean $= \dfrac{1}{5}(1 + 3 + 5 + 9 + 13) = 6.2$

Median = the middle value = 5

The median was the middle value of the numbers as they were presented in the question only because the data was already arranged in order of magnitude.

b. These numbers are ten times those in *a.* Hence their mean

$$= 10 \times 6.2 = 62$$

c. The mean of these numbers $= 51 + 6.2 = 57.2$

3.2a. With 10 as a class interval, the modal class is 109.5 to under 119.5 and the midpoint $= 109.5 + 5 = 114.5$. With 20 as a class interval, the modal class is 89.5 to under 109.5 and the mid point $= 89.5 + 10 = 99.5$

b. We required the $\dfrac{110 + 1}{2}$ or $55\frac{1}{2}$th hypothetical value. With 10 as a class interval there are 55 values below 109.5. The 55th value is thus in the top of the range 99.5 — under 109.5 and the 56th value in the bottom of the range 109.5 — under 119.5. Hence the $55\frac{1}{2}$th value or median is 109.5. The same result follows with 20 as class interval.

c.(i)

Class	(1) Mid point	(2) Frequency	(1) × (2)
159.5 — under 169.5	164.5	1	164.5
149.5 — under 159.5	154.5	2	309.0
139.5 — under 149.5	144.5	6	867.0
129.5 — under 139.5	134.5	10	1345.0
119.5 — under 129.5	124.5	16	1992.0
109.5 — under 119.5	114.5	20	2290.0
99.5 — under 109.5	104.5	19	1985.5
89.5 — under 99.5	94.5	18	1701.0
79.5 — under 89.5	84.5	12	1014.0
69.5 — under 79.5	74.5	3	223.5
59.5 — under 69.5	64.5	2	129.0
49.5 — under 59.5	54.5	1	54.5
		110	12075.0

$$\therefore \text{Arithmetic mean} = \frac{12075}{110} = 109.8$$

Class	(1) Mid point	(2) Frequency	(1) × (2)
149.5 — under 169.5	159.5	3	478.5
129.5 — under 149.5	139.5	16	2232.0
109.5 — under 129.5	119.5	36	4302.0
89.5 — under 109.5	99.5	37	3681.5
69.5 — under 89.5	79.5	15	1192.5
49.5 — under 69.5	59.5	3	178.5
		110	12065.0

$$\therefore \text{Arithmetic mean} = \frac{12065}{110} = 109.7$$

The slightly different values obtained for the arithmetic mean result from the slightly different groupings.

c.(ii)   Select 104.5 as arbitrary origin and 10 as unit.

Class	(1) Mid point	(2) Deviation	(3) Frequency	(2) $\times$ (3)
159.5 — under 169.5	164.5	6	1	6
149.5 — under 159.5	154.5	5	2	10
139.5 — under 149.5	144.5	4	6	24
129.5 — under 139.5	134.5	3	10	30
119.5 — under 129.5	124.5	2	16	32
109.5 — under 119.5	114.5	1	20	20
99.5 — under 109.5	104.5	0	19	0
89.5 — under 99.5	94.5	—1	18	—18
79.5 — under 89.5	84.5	—2	12	—24
69.5 — under 79.5	74.5	—3	3	— 9
59.5 — under 69.5	64.5	—4	2	— 8
49.5 — under 59.5	54.5	—5	1	— 5
			110	58

$$\therefore \text{Arithmetic mean} = 104.5 + \frac{58}{110} \times 10 = 109.8 \text{ (as before)}$$

Select 99.5 as origin and 20 as unit.

Class	(1) Mid point	(2) Deviation	(3) Frequency	(2) $\times$ (3)
149.5 — under 169.5	159.5	3	3	9
129.5 — under 149.5	139.5	2	16	32
109.5 — under 129.5	119.5	1	36	36
89.5 — under 109.5	99.5	0	37	0
69.5 — under 89.5	79.5	—1	15	—15
49.5 — under 69.5	59.5	—2	3	— 6
			110	56

$$\therefore \text{Arithmetic mean} = 99.5 + \frac{56}{110} \times 20 = 109.7 \text{ (as before)}$$

3.3 The total of the heights (by addition) $\quad = 3519$ cm
The number of jumps $\quad = 60$

$\therefore$ the arithmetic mean of the heights $\quad = 58\dfrac{39}{60}$ cm

If we use 55 as an arbitrary origin the heights become

$$
\begin{array}{rrrrrr}
+12 & + 5 & - 2 & +12 & + 8 & +16 \\
+ 8 & + 1 & - 6 & + 8 & + 9 & +15 \\
- 6 & - 2 & + 5 & + 2 & + 6 & +10 \\
+ 2 & - 2 & - 5 & + 5 & - 5 & - 4 \\
+12 & +13 & + 3 & - 1 & + 6 & +11 \\
+ 3 & - 1 & - 5 & +10 & +13 & - 7 \\
- 9 & +17 & +11 & +14 & +10 & + 3 \\
+15 & - 9 & + 7 & - 4 & - 8 & + 7 \\
- 7 & + 4 & + 3 & + 7 & 0 & - 3 \\
- 3 & + 1 & + 8 & + 4 & + 1 & + 1 \\
\end{array}
$$

The total of these heights $= 219$

$\therefore$ the arithmetic mean $= 55 + \dfrac{219}{60} = 58\dfrac{39}{60}$ cm as before.

From the solution to *2.1* it can be seen that the 30th and 31st heights are 58 and 59 cm respectively. Hence the median $= 58\frac{1}{2}$ cm. The heights 56, 58 and 63 cm each occur 4 times and this is the highest frequency. Hence each of these is the modal height.

3.4 Arithmetic mean $= \dfrac{1}{5}(175{,}264 + 4685 - 14617 - 7245 - 1654)$

$\qquad\qquad\qquad\quad = \$31{,}286.6$ profit

Median value $\qquad = \$1{,}654$ loss.

These vary greatly. Neither is representative. In fact it is not meaningful to represent such a varying collection of figures by an average figure.

4.1 Total score for 1st cricketer $=$ 385 for 10 innings
$\qquad\therefore$ His batting average $=$ 38.5 runs
$\qquad$ Total score for 2nd cricketer $=$ 385 for 10 innings
$\qquad\therefore$ His batting average $=$ 38.5 runs also

They have the same average; but the second is the more consistent as all his scores lie between 32 and 44.

The first batsman scored 142 but twice failed to reach double figures. Some indication of the respective variations is given by the standard deviations which are 41.8 and 4.0 respectively.

4.2a. Presumably the measurements are given to the nearest centimetre. The class intervals should be defined as

$2\frac{1}{2}$ cm to under $3\frac{1}{2}$ cm , $3\frac{1}{2}$ cm to under $4\frac{1}{2}$ cm , etc.

b. There can be no wage between \$11.99 and \$12.00 because there is no coin less than one cent. Hence the class intervals in dollars should be

10.00 to 11.99, 12.00 to 13.99, etc.

4.3a. The mean $= \dfrac{45}{5} = 9.$

The deviations from the mean are —8, —6, —2, +4, +12.

$$\therefore s = \sqrt{\frac{\Sigma(x - \bar{x})^2}{n}} = \sqrt{\frac{264}{5}} = 7.27.$$

b. Mean deviation $= \dfrac{\Sigma|x - \bar{x}|}{n} = \dfrac{32}{5} = 6.4.$

c. These figures are each 10 times those in *a.* Hence their mean and standard deviation will be 10 times the mean and standard deviation of the figures in *a.* It is in effect merely a change of units, similar to changing yards to feet.

$\therefore$ Standard deviation $= 72.7.$

d. If 7 is added to each figure we are merely changing the origin. The variation is unaltered and hence the standard deviation is unaltered and is therefore 72.7.

4.4 We must first calculate the standard deviation. Select 55 as arbitrary origin and 5 marks as unit of deviation.

Mark	Deviation (d)	Frequency (f)	$f \times d$	$f \times d^2$
45	—2	2	— 4	8
50	—1	3	— 3	3
55	0	20	0	0
60	+1	16	+16	16
65	+2	6	+12	24
70	+3	1	+ 3	9
Totals		48	+24	60

Mean value of $d = \dfrac{24}{48} = \dfrac{1}{2}$

$$s = \sqrt{\frac{\Sigma fd^2}{\Sigma f} - \left(\frac{\Sigma fd}{\Sigma f}\right)^2} = \sqrt{\frac{60}{48} - \left(\frac{24}{48}\right)^2} = 1$$

The mean is thus $\frac{1}{2}$ a unit of d above the arbitrary origin of 55. The mean is therefore $57\frac{1}{2}$. The SD is 1 unit of d, namely 5 marks.
The marks in standard scores are therefore

$$\frac{45 - 57\frac{1}{2}}{5}, \quad \frac{50 - 57\frac{1}{2}}{5} \text{ etc.}$$

$$\text{i.e. } -2.5, -1.5, -0.5, +0.5, +1.5, +2.5.$$

4.5a. 110,
115, 118,
120, 122, 124, 124, 125, 126, 128,.128, 128,
130, 131, 132, 134, 136, 138,
140, 142.

$$b. \ \overline{x} = \frac{2551}{20} = 127.55 \text{ mm.}$$

The 10th and 11th items are both 128. Hence the median (or $10\frac{1}{2}$th item) = 128 mm.
The lower quartile (or $5\frac{1}{4}$th item) lies between 122 and 124 and hence may be taken as $122\frac{1}{2}$ (say).
The upper quartile (or $15\frac{3}{4}$th item) lies between 132 and 134 and hence may be taken as $133\frac{1}{2}$ (say).
  c. The range $= 142 - 110 = 32$ mm
Semi-interquartile range $= \frac{1}{2}(133\frac{1}{2} - 122\frac{1}{2}) = 5.5$ mm

$$\text{Standard deviation } = \sqrt{\frac{\Sigma x^2}{n} - \overline{x}^2}$$

$$= \sqrt{\frac{326,683}{20} - (127.55)^2}$$

$$= \quad 8.07 \text{ mm}$$

To find the mean deviation, we total the deviations (*irrespective of sign*) of all values from their mean 127.55.
These deviations are

4.45,	12.55,	5.55,	14.45,	3.55,	8.45,	9.55,
0.45,	0.45,	2.45,	1.55,	2.55,	6.45,	0.45,
7.55,	12.45,	3.55,	10.45,	17.55,	3.55,	

totalling 128.00.

$$\therefore \text{ Mean deviation } = \frac{128.00}{20} = 6.4 \text{ mm.}$$

d. As pressure is a continuous variable, the readings are presumably taken to the nearest unit. That is, the actual value of blood pressure resulting in a reading of 110 could be anywhere in the range $109\frac{1}{2}$ to $110\frac{1}{2}$. The values 110 to 114 inclusive could in reality be the range $109\frac{1}{2}$ to $114\frac{1}{2}$. Hence our frequency table is the first two columns of the following table: (The other columns are required later.)

Class interval	Frequency (f)	Mid point (x)	$f \times x$	$f \times x^2$
$109\frac{1}{2}$ — under $114\frac{1}{2}$	1	112	112	12544
$114\frac{1}{2}$ — under $119\frac{1}{2}$	2	117	234	27378
$119\frac{1}{2}$ — under $124\frac{1}{2}$	4	122	488	59536
$124\frac{1}{2}$ — under $129\frac{1}{2}$	5	127	635	80645
$129\frac{1}{2}$ — under $134\frac{1}{2}$	4	132	528	69696
$134\frac{1}{2}$ — under $139\frac{1}{2}$	2	137	274	37538
$139\frac{1}{2}$ — under $144\frac{1}{2}$	2	142	284	40328
Totals	20		2555	327665

e.

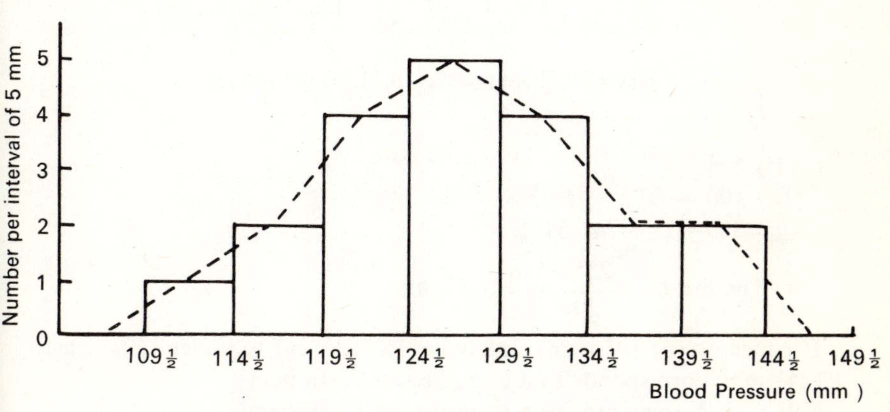

Histogram for blood pressure of 20 healthy persons

f. These may be expressed as the percentages of cases in each class interval, namely:

$$5, 10, 20, 25, 20, 10, 10 \text{ (Total} = 100).$$

g. One suitable approach would be to use the ogive drawn with percentage relative frequencies as this enables the required percentages to be read directly. The percentage relative frequencies are:

$$5, 15. 35, 60, 80, 90, 100.$$

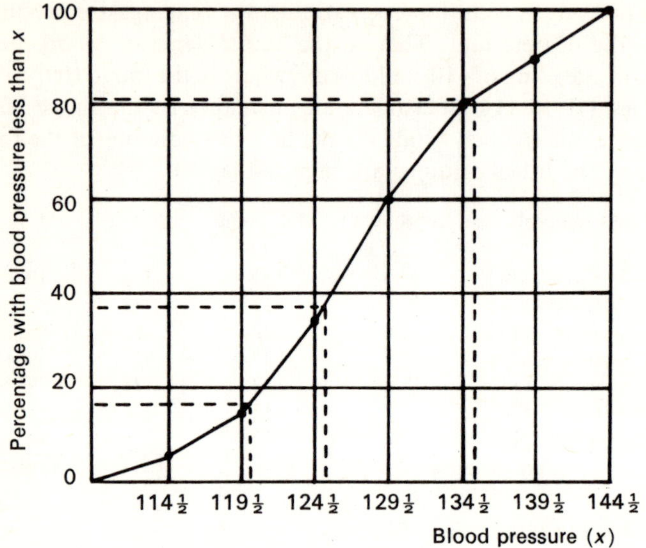

Ogive for blood pressure of 20 healthy persons

(*i*) 81%
(*ii*) 100 — 37.5 = 62.5%
(*iii*) 81 — 17   = 64%.

h. The mean $= \dfrac{2555}{20} = 127.75$ mm.

The median (or 10½ item) clearly lies in the '124½ to under 129½' class.
124½ mm. corresponds to a hypothetical 7½th item;
129½ mm. corresponds to a hypothetical 12½th item.

$$\therefore \text{Median} = 124\tfrac{1}{2} + \frac{10\tfrac{1}{2} - 7\tfrac{1}{2}}{12\tfrac{1}{2} - 7\tfrac{1}{2}} \times 5 = 127\tfrac{1}{2} \text{ mm}$$

Similarly,

$$\text{Lower quartile} = 119\tfrac{1}{2} + \frac{5\tfrac{1}{4} - 3\tfrac{1}{2}}{7\tfrac{1}{2} - 3\tfrac{1}{2}} \times 5 = 121.7 \text{ mm}$$

$$\text{Upper quartile} = 129\tfrac{1}{2} + \frac{15\tfrac{3}{4} - 12\tfrac{1}{2}}{16\tfrac{1}{2} - 12\tfrac{1}{2}} \times 5 = 133.6 \text{ mm}.$$

i.

$$s = \sqrt{\frac{\Sigma fx^2}{\Sigma f} - \left(\frac{\Sigma fx}{\Sigma f}\right)^2}$$

$$\frac{\Sigma fx^2}{\Sigma f} = \frac{327{,}665}{20} = 16{,}383.25$$

$$\left(\frac{\Sigma fx}{\Sigma f}\right)^2 = \left(\frac{2555}{20}\right)^2 = 16320.06$$

$$\therefore s = \sqrt{63.19} = 7.95 \text{ mm.}$$

This value differs slightly from the value obtained in *c.* for *s* from the raw data (8.07 mm). The difference is due to the loss of accuracy on grouping.

j.   From *b.* and *c.*   Mean $=$ 127.55 mm

SD $=$ 8.07 mm

We are therefore required to find the percentage of cases lying within the range 119.48 to 135.62 mm

From *a.* this percentage $= \dfrac{13}{20} \times 100 = 65\%$.

4.6

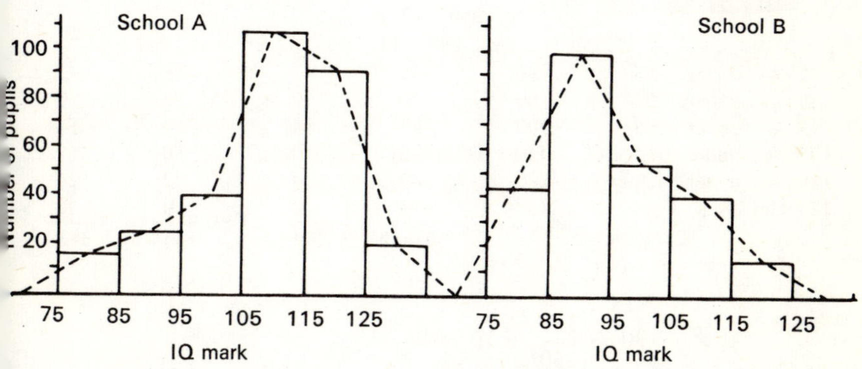

IQ marks for pupils at two schools

a.

School A

IQ Mark	Mid point (x)	Deviation from 110 with 10 as unit (d)	Frequency (f)	$f \times d$	$f \times d^2$
75 — under 85	80	—3	15	—45	135
85 — under 95	90	—2	25	—50	100
95 — under 105	100	—1	40	—40	40
105 — under 115	110	0	108	0	0
115 — under 125	120	+1	92	92	92
125 and over	130	+2	20	40	80
Totals			300	—3	447

$$\overline{x} = 110 - \frac{3}{300} \times 10 = 109.9$$

$$s = 10 \sqrt{\frac{447}{300} - \left(\frac{-3}{300}\right)^2} = 12.2$$

School B

IQ Mark	Mid point (x)	Deviation from 90 with 10 as unit (d)	Frequency (f)	$f \times d$	$f \times d^2$
75 — under 85	80	—1	43	—43	43
85 — under 95	90	0	99	0	0
95 — under 105	100	+1	54	54	54
105 — under 115	110	+2	40	80	160
115 — under 125	120	+3	14	42	126
125 and over	130	+4	0	0	0
Totals			250	133	383

$$\overline{x} = 90 + \frac{133}{250} \times 10 = 95.3$$

$$s = 10 \sqrt{\frac{383}{250} - \left(\frac{133}{250}\right)^2}$$
$$= 11.2$$

b.

School A

$$\text{Median IQ (150}\tfrac{1}{2}\text{th pupil)} \quad = 105 + \frac{150\tfrac{1}{2} - 80\tfrac{1}{2}}{188\tfrac{1}{2} - 80\tfrac{1}{2}} \times 10 = 111.5$$

$$\text{Upper quartile (225}\tfrac{3}{4}\text{th pupil)} = 115 + \frac{225\tfrac{3}{4} - 188\tfrac{1}{2}}{280\tfrac{1}{2} - 188\tfrac{1}{2}} \times 10 = 119.05$$

$$\text{Lower quartile (75}\tfrac{1}{4}\text{th pupil)} \quad = 95 + \frac{75\tfrac{1}{4} - 40\tfrac{1}{2}}{80\tfrac{1}{2} - 40\tfrac{1}{2}} \times 10 = 103.69$$

$$\text{Interquartile range} \quad\quad = 119.05 - 103.69 \quad\quad = 15.4$$

School B

$$\text{Median IQ (125}\tfrac{1}{2}\text{th pupil)} \quad = 85 + \frac{125\tfrac{1}{2} - 43\tfrac{1}{2}}{142\tfrac{1}{2} - 43\tfrac{1}{2}} \times 10 = 93.3$$

$$\text{Upper quartile (188}\tfrac{1}{4}\text{th pupil)} = 95 + \frac{188\tfrac{1}{4} - 142\tfrac{1}{2}}{196\tfrac{1}{2} - 142\tfrac{1}{2}} \times 10 = 103.47$$

$$\text{Lower quartile (62}\tfrac{3}{4}\text{th pupil)} \quad = 85 + \frac{62\tfrac{3}{4} - 43\tfrac{1}{2}}{142\tfrac{1}{2} - 43\tfrac{1}{2}} \times 10 = 86.94$$

$$\text{Interquartile range} \quad\quad = 103.47 - 86.94 \quad\quad = 16.5$$

c.

IQ Mark	Number of pupils of both schools combined
75 — under 85	58
85 — under 95	124
95 — under 105	94
105 — under 115	148
115 — under 125	106
125 and over	20
Total	= 550

We require the 80th percentile, i.e. the IQ of the $\dfrac{551}{100} \times 80$ or 440.8th pupil. Adopting a similar approach to that used on page 38 for calculating quartiles from grouped data we obtain for the 80th percentile

$$115 + \frac{440.8 - 424\tfrac{1}{2}}{530\tfrac{1}{2} - 424\tfrac{1}{2}} \times 10 = 116.5.$$

d. The mean IQ of pupils in schools A, B and C combined

$$= \frac{300 \times 109.9 + 250 \times 95.3 + 450 \times 106}{300 + 250 + 450}$$

$$= \frac{104495}{1000}$$

$$\doteqdot 104.5.$$

5.1 There are 16 possible families, which can easily be seen from the following 'tree' of possibilities:

1st child			B				G	
2nd child		B		G		B		G
3rd child	B	G	B	G	B	G	B	G
4th child	BG	BG	BG	BG	BG	BG	BG	BG

a. The possibilities are BGGG, GBGG, GGBG, GGGB

i.e. 4 in number. $\therefore$ Probability $= \dfrac{4}{16}$ or $\dfrac{1}{4}$

b. The possibilities are BBGG, BGBG, BGGB, GBBG, GBGB, GGBB

i.e. 6 in number. $\therefore$ Probability $= \dfrac{6}{16}$ or $\dfrac{3}{8}$

c. The only possibility is BBGG. $\therefore$ Probability $= \dfrac{1}{16}$

d. The only possibility is BGGG. $\therefore$ Probability $= \dfrac{1}{16}$

5.2   The chance A will die in a year $= \dfrac{4}{100}$ or .04

$\therefore$ The chance A will live for a year $= \dfrac{96}{100}$ or .96

The chance B will die in a year $= \dfrac{3}{100}$ or .03

$\therefore$ The chance B will live for a year $= \dfrac{97}{100}$ or .97

a. These are independent events and we require *both* to happen
   $\therefore$ Probability $= .04 \times .97 = .0388$
b. Similarly, the probability in this case $= .04 \times .03 = .0012$

c. Similarly, the probability in this case $= .96 \times .97 = .9312$

d. Either one of the two following possibilities *must* occur:

Either *(i)* A and B will both live a year,

or   *(ii)* at least one of A or B will die in a year.

The probability of *(i)* we worked out in *c.* to be .9312.

$\therefore$ the probability of *(ii)* $= 1 - .9312 = .0688$.

5.3 This is not a satisfactory survey because the results are almost certain to be biased.

If the questions are to be answered in respect of the students themselves, one would hope that their television watching was smaller in quantity than the average person. If the type of programme being watched was the matter of study, in view of their age and educational level, this too would not be fairly represented by university students.

Even if the questions are to be answered in respect of the student's family, bias is still likely to occur.

One correct approach would be to select a large group at random in some suitable way.

5.4 The frequencies of 0, 1, 2, 3 and 4 tails are 2, 5, 6, 5, 2 respectively.

$$\text{Hence } \Sigma f = 2 + 5 + 6 + 5 + 2 = 20$$
$$\Sigma fx = 2 \times 0 + 5 \times 1 + 6 \times 2 + 5 \times 3 + 2 \times 4 = 40$$
$$\Sigma fx^2 = 2 \times 0 + 5 \times 1 + 6 \times 4 + 5 \times 9 + 2 \times 16 = 106$$

$$\text{Therefore } \bar{x} = \frac{40}{20} = 2, \text{ and } s = \sqrt{\frac{106}{20} - \left(\frac{40}{20}\right)^2} = 1.14$$

$$\textbf{5.5} \quad \Sigma f = 0 + 7 + 9 + 11 + 3 + 2 = 32$$
$$\Sigma fx = 1 \times 0 + 7 \times 1 + 9 \times 2 + 11 \times 3 + 3 \times 4 + 2 \times 5 = 80$$
$$\Sigma fx^2 = 1 \times 0 + 7 \times 1 + 9 \times 4 + 11 \times 9 + 3 \times 16 + 2 \times 25 = 240$$

$$\text{Hence } \bar{x} = \frac{80}{32} = 2.5$$

$$s = \sqrt{\frac{240}{32} - \left(\frac{80}{32}\right)^2} = 1.118$$

5.6 *a.* $0.81590 - 0.30865 = 0.50725$

b. 0.75794

c. 0.69135

6.1a. One hundred sixty-three cm is a standard score of $\dfrac{163 - 175}{9}$ or

— 1.33.

The required probability is therefore that of obtaining a standard score of less than —1.33.

∴ Probability $= .5 — .4082 = .0918$ or (say) *nine in one hundred.*

  b. This is the probability of obtaining a standard score between $\dfrac{163 - 175}{9}$ and $\dfrac{178 - 175}{9}$, i.e. between —1.33 and +0.33.

$$= .4082 + .1293 \text{ (from normal tables)}$$
$$= .5375 \text{ say } 54 \text{ in } 100.$$

6.2 98 is a standard score of —0.2. The proportion below this figure is therefore

$$0.5 — 0.0793 = .4207$$

The proportion with IQs between 98 and 110 is the proportion with a standard score between —0.2 and +1.0. This proportion is

$$0.0793 + 0.3413 = .4206$$

The proportion with IQs above 110 is the proportion with a standard score exceeding 1.0 which equals

$$0.5 — 0.3413 = .1587$$

(Students cannot be in two of these groups; that is, the groups are mutually exclusive. Hence the probability that they are in one of the three groups is the sum of the three probabilities. But they must lie in one of the three groups, i.e. the probability of this equals unity. Hence the sum of the three proportions must total unity, as they do.)

Since there are 10,000 students altogether the numbers in the three groups would be

a.   4207

b.   4206

c.   1587.

6.3 Normal distribution tables show that 20% of a normal population have a standard score exceeding 0.84. To achieve an A a student should therefore obtain a standard score of 0.84 or more, i.e. a mark of .84 $\times$ 20 above the mean of 55. The A mark would therefore be $55 + .84 \times 20 = 72.$

6.4 Since $1\frac{1}{2}\%$ of bolts are to be rejected, we shall assume that $\frac{3}{4}\%$ will be rejected because they are too long and $\frac{3}{4}\%$ because they are too short. If $\frac{3}{4}\%$ or 0.0075 of bolts are too long then $0.5 - .0075$ or 0.4925 of bolts must be longer than the mean *but* acceptable.

The standard score which has 0.4925 of cases lying between it and the mean is found by entering the normal tables in reverse fashion. The result is a standard score of 2.43. Since the SD is 0.2 cm the maximum limit is $10 + 2.43 \times 0.2$ or 10.486 cm and the minimum limit $10 - 2.43 \times 0.2$ or 9.514 cm.

6.5a. The required chance is the chance of a standard score between $-\frac{1}{4}$ and $+\frac{1}{4}$ which equals $.0987 + .0987 = .1974$.

 b. The average contents of 100 bottles would be normally distributed with a mean of 80 cc and a standard deviation of $\dfrac{4}{\sqrt{100}}$ or 0.4 cc.

The required probability is therefore the chance of a standard score between $-\dfrac{1}{.04}$ and $+\dfrac{1}{.04}$, i.e. between -2.5 and $+2.5$.

$\therefore$ The chance $= 0.4938 + 0.4938 = 0.9876$.

6.6 x is normally distributed with a mean of 30 and a standard deviation of 8. The mean of a sample of 16 values of x is therefore normally distributed with a mean of 30 and a standard deviation of $\dfrac{8}{\sqrt{16}}$ or 2.

a. The probability that a particular sample had a mean less than 32
  $=$ probability of a standard score less than $+ 1.0$
  $= 0.5 + 0.3413 = 0.84$.

b. Probability of a sample mean greater than 36
  $=$ probability of a standard score greater than 3.0
  $= 0.5 - 0.4987 = .0013$ (say) 1 in 1000.

c. Probability of a sample mean greater than 28 is the probability of a standard score exceeding -1.0
  $= 0.3413 + 0.5 = .8413$ (say) 84 in 100.

d. Probability of a sample mean less than 25 is the probability of a standard score less than -2.5
  $= 0.5 - 0.4938 = .0062$ (say) 6 in 1000.

e. Probability of a sample mean between 33 and 34 is the probability of a standard score between $+ 1.5$ and $+ 2.0$
  $= 0.4772 - 0.4332 = .044$.

7.1 Individual marks over a long period have a mean of 60 and standard deviation 10. If we had grouped these marks at random in batches of 72 and found the mean of each batch, then the set of figures so obtained would have a mean of 60 and standard deviation of $\dfrac{10}{\sqrt{72}}$ or 1.18. Even if the individual marks are not normally distributed, we may assume from the Central Limit Theorem that in samples of this size the mean is normally distributed. With samples of 72, the mean is thus distributed as $N(60, 1.18)$.

This particular school has produced a sample with a mean of 64. This is a deviation of $\dfrac{4}{1.18}$ or 3.39 standard deviations. The probability of a deviation of this amount or more due to chance alone is much less than 1 in 1000. This particular group of students can claim to have produced a superior result in mathematics.

7.2 Assume that there is no real increase, and that any variation is due to chance. That is, we assume that this is a random sample of 16 from a population that is normally distributed with mean 600 kg and standard deviation 25 kg. The mean of samples of size 16 from such a population would be normally distributed with mean 600 and standard deviation $\dfrac{25}{\sqrt{16}}$ or 6.25.

This particular sample has $\overline{x} = \dfrac{10037}{16} = 627.3$.

This is $\dfrac{27.3}{6.25}$ or 4.37 standard deviations above the mean. We cannot determine the chance of a deviation this amount or greater from our tables but it is clearly much less than 1 in 1000. Hence we may reasonably conclude that the new process has increased the breaking strength.

7.3 The analysis proceeds as follows:
(*i*) Assume that there is no real difference in deterioration time between the two varieties and that these are therefore two samples from the same population.
(*ii*) On our assumption we have two samples of 50 from the same population and the difference of the means of the two samples would therefore be distributed with mean zero and standard deviation

$$\sqrt{\frac{9^2}{50} + \frac{9^2}{50}} = 1.8$$

In view of the size of these samples, the difference between the means would be nearly normally distributed.

(*iii*) In this particular case the difference
$$= 93.6 - 90.9 = 2.7.$$

This equals $\dfrac{2.7}{1.8}$ or $1\frac{1}{2}$ standard deviations.

(*iv*) A difference of this amount or greater can occur reasonably often by chance. Hence we cannot conclude on this evidence that there is a real difference in deterioration time between the two varieties of paint.

7.4a. Assume that there is no real difference between his product and the national product and that the apparent variation is due to chance. That is, we assume that this is a sample of 64 ears from a population which is normally distributed with mean 40 mm and standard deviation 2.04 mm.

The means of samples of size 64 from such a population would be normally distributed with mean 40 mm. and standard deviation $\dfrac{2.04}{\sqrt{64}}$

or 0.255 mm. His sample is therefore $\dfrac{40.7 - 40.0}{0.255}$ or 2.75 standard deviations above the mean. The chance of a deviation of this size or greater is $0.5 - 0.497 = 0.003$. The probability of this being a random sample from the population is therefore small. He could, on this evidence, reasonably claim to have a product larger than the national average.

b. Assume that there is no real difference between the mean diameters of their product. That is, that the difference which occurred is due to chance.

The standard deviation of the difference in the means of samples of these sizes is

$$\sqrt{\frac{2.04^2}{64} + \frac{2.04^2}{49}} = 0.387$$

The actual difference between the means is 0.4 mm or little more than one standard deviation. This could reasonably occur by chance and hence on this evidence there is no clear justification for the claim.

8.1 Mean $=$ $1000 \times \frac{1}{2}$ $= 500$ heads

SD $= \sqrt{1000 \times \frac{1}{2} \times \frac{1}{2}}$ $= 15.81$ heads.

8.2 Mean $=$ $1200 \times \frac{1}{6}$ $= 200$ fours

SD $= \sqrt{1200 \times \frac{1}{6} \times \frac{5}{6}}$ $= 12.9$ fours.

8.3 If we repeatedly chose classes of 200 at random we would find that the number of passes would vary around a mean of 100 with a standard deviation of $\sqrt{200 \times \frac{1}{2} \times \frac{1}{2}}$ or $\sqrt{50}$ which equals 7.071.

We require to know how frequently among these classes we are likely to find one in which 110 or more pass i.e. 10 more than the mean. Using normal distribution tables as an approximation and applying the continuity correction, we obtain for this probability the probability under a normal distribution of a standard score of $\dfrac{9\frac{1}{2}}{7.071}$ i.e. 1.34 or more

$$= 0.5 - 0.4099$$
$$= 0.0901 \text{ or about 9 in 100.}$$

8.4 If we repeatedly observe randomly chosen schools of 300 pupils we would find that the number medically unfit in these schools would vary with a mean of 90 (i.e. 30% of 300) and standard deviation of $\sqrt{300 \times .3 \times .7}$ or $\sqrt{63}$ which equals 7.937.

The probability of finding a school with 105 or more medically unfit is the probability of a value of 15 or more above the mean.

Using normal distribution tables as an approximation and applying the continuity correction, we obtain for this probability the probability under a normal distribution of a standard score of $\dfrac{14\frac{1}{2}}{7.937}$ i.e. 1.83 or more

$$= 0.5 - 0.4664$$
$$= 0.0336 \text{ or about 1 in 30.}$$

8.5 The number of washing machines breaking down in randomly chosen groups of 100 varies with a mean of 10 and a SD of

$$\sqrt{100 \times 0.1 \times 0.9} = 3.$$

a. The probability of having 18 or more break down
= the probability of a value 8 or more above the mean.
Using normal distribution tables as an approximation and applying
the continuity correction we obtain for this probability the probability
under a normal distribution of a value $\dfrac{7\frac{1}{2}}{3}$ or $2\frac{1}{2}$ standard deviations or

more above the mean
$$= 0.5 - 0.4938 = 0.0062 \text{ or 6 in } 1000.$$
b. The probability of at least 5 but not more than 15 from normal
tables and applying the continuity correction

= the probability of a standard score between $\dfrac{-5\frac{1}{2}}{3}$ or -1.83 and $\dfrac{+5\frac{1}{2}}{3}$

or $+1.83$
$= 0.4664 + 0.4664$
$= 0.9328.$
The chance is over 93 in 100.

8.6 Assume that the new treatment is not really more successful and
that the apparently better result is due to chance.
The population value of π is 0.2. Random samples of 200 would have
$$\mu = \quad 200 \times 0.2 \qquad = 40$$
$$\sigma = \sqrt{200 \times 0.2 \times 0.8} \quad = 5.657$$
Using the normal distribution as an approximation to the binomial
and applying the continuity correction, we obtain for the chance of 60
or more successes the probability of a standard score exceeding $\dfrac{19\frac{1}{2}}{5.657}$

This is a standard score greater than 3 and the probability of obtaining
such a result by chance is less than 1 in 1000. This is most unlikely and
we may conclude that the new treatment is more successful.
We have used a one-tailed test because of the definite suggestion that
the new treatment was more successful.

8.7 Assume that this year's class is a random sample of 400 students
from a large population in which 75% pass and that the poorer result
is purely due to chance.
Random samples of 400 from this population would have
$$\mu = \quad 400 \times \tfrac{3}{4} \qquad = 300$$
$$\sigma = \sqrt{400 \times \tfrac{3}{4} \times \tfrac{1}{4}} \quad = 8.66$$

Our particular sample has produced a result which is 10 below the mean.

We shall use a two-tailed approach as the professor appears to be studying yearly variations. At the 5% level of significance with a normal distribution we would expect a standard score lying between —1.96 and + 1.96.

Using the normal distribution as an approximation to the binomial, and ignoring the continuity correction we obtain for the range of acceptance of the hypothesis

$$300 - 1.96 \times 8.66 \text{ to } 300 + 1.96 \times 8.66$$

i.e. from 283 to 317.

The number of passes this year is well within this range and hence there is no reason to suspect the hypothesis. The poorer result this year could reasonably have happened by chance.

10.1 The analysis proceeds as follows:

(*i*) Assume that this group is no better than the students as a whole. That is, we assume these results are a randon sample from a normal population with mean 62 and unknown standard deviation.

(*ii*) For the sample

$$\text{Mean} = \frac{1091}{17} = 64.176$$

$$\text{Standard deviation} = \sqrt{\frac{70203}{17} - 64.176^2} = 3.32$$

Therefore

$$t = \frac{64.176 - 62}{3.32}\sqrt{16}$$

$$= 2.62 \text{ with 16 degrees of freedom.}$$

(*iii*) With 16 degrees of freedom, tables of the *t* distribution show that a one-tailed test at the 5% significance level, values of *t* up to 1.75 may be expected.

(*iv*) We conclude that this group is significantly better than the students as a whole.

10.2 The analysis proceeds as follows:

(*i*) Assume that the reprimand is not justified and that the difference is due to chance. Assume also that the lengths are normally distributed.

(*ii*) For the lengths produced by the apprentice

$$t = \frac{22 - 18}{6} \sqrt{36}$$

$$= 4 \text{ with 36 degrees of freedom.}$$

(*iii*) This difference is highly significant. Hence our assumption is almost certainly wrong and we conclude that the reprimand is justified.

10.3 The analysis proceeds as follows:

(*i*) Assume that conditions have no effect on marks. That is, assume that these are two samples from the same population.

(*ii*) The difference between the means will be distributed normally (in view of the size of the samples) with zero mean and unknown standard deviation. Hence the *t* distribution applies.

$$t = \frac{93.6 - 90.9}{\sqrt{\dfrac{50 \times 8^2 + 50 \times 10^2}{50 + 50 - 2}}} \sqrt{\frac{50 \times 50}{100}} = 1.47$$

Degrees of freedom $= 50 + 50 - 2 = 98$.

(*iii*) At the 5% level of significance with 98 degrees of freedom values of *t* up to 1.99 are expected with a two tailed test.

(*iv*) The difference recorded could well have occurred by chance and hence on this evidence we cannot conclude that quiet conditions make for better performance.

10.4 If the first two sentences in the question had been omitted we would have had two samples of 12 and we would have tested the difference between the means of these two samples treating the samples as being from a normal population with unknown variance.

However, the first two sentences make it clear that the values are 'paired' — 52 and 42 being the results of the two types of painting *applied to the one house*. This is to make sure that the experiment is a fair one and to eliminate differences which might occur if we compared painting performed at different distances from sea air, from smoky industrial areas etc. Hence we compare each value with its partner. To do this we calculate the differences between 'two undercoats' and 'one undercoat'. These differences are

$$+10, \ +3, \ +3, \ +22, \ +18, \ -2, \ +28, \ +12, \ +8, \ -10, \ +4, \ -6$$

The analysis then proceeds as follows:

(*i*) Assume that the distribution of lasting effects is normal and that there is no difference in the lasting effects of the different coatings. That is, we assume that the differences obtained are from a population whose mean is zero.

(*ii*) For the sample of 12 differences we obtain

$$\bar{x} = \frac{90}{12} = 7.5$$

$$s = \sqrt{\frac{2074}{12} - 7.5^2} = 10.80$$

$$t = \frac{7.5 - 0}{10.8}\sqrt{11} = 2.30$$

(*iii*) At the 5% level of significance with 11 degrees of freedom values of t up to 1.80 may be expected with a one tailed test.

(We use a one tailed test because of the definite suggestion that the second undercoat produces a more lasting result.)

(*iv*) The value produced by this sample is therefore unlikely to have occurred by chance and we conclude that the second undercoat evidently produces a more lasting result.

11.1 Assume that visits are uniformly spread over the day. The actual and expected numbers are then as follows:

											Total
Actual	14	8	4	19	12	11	14	6	8	4	100
Expected	10	10	10	10	10	10	10	10	10	10	100

$$\chi^2 = \frac{4^2}{10} + \frac{(-2)^2}{10} + \cdots\cdots\cdots\cdots + \frac{(-2)^2}{10} + \frac{(-6)^2}{10}$$

$$= 21.4$$

With a total of 100 visits, if we are given the number during the first 9 hours, the number during the tenth hour can be calculated by subtraction. Therefore there are 9 independent values or 9 degrees of freedom. The chance of obtaining this value of χ^2 or higher with 9 degrees of freedom is about 1 in 100. Hence we conclude that visits are not spread uniformly over the 10 hours.

11.2 Assume the mixture and the water have in reality the same effect.
The expected values are then as shown in brackets.

	Better	Worse	No change	Total
Water	110 (117.5)	40 (35)	50 (47.5)	200
Mixture	125 (117.5)	30 (35)	45 (47.5)	200
Total	235	70	95	400

$$\chi^2 = \frac{(110 - 117.5)^2}{117.5} + \frac{(40 - 35)^2}{35} + \ldots + \frac{(45 - 47.5)^2}{47.5}$$

$$= 2.65$$

Once any two values are filled in, all other values are fixed by the
marginal totals. Hence there are 2 degrees of freedom. At the 5%
level of significance with 2 degrees of freedom values of χ^2 up to 5.99
are expected. As a value as large as that observed could quite
easily have arisen by chance we conclude that there is no evidence
from this experiment that the mixture has significantly better results
than water.

11.3 Assume colour of hair and colour of eye are independent of one
another. On this assumption the expected numbers would be as
indicated in brackets.

Eye	Hair				Total
	Fair	Brown	Black	Red	
Blue	1768 (1169)	807 (1088)	189 (506)	47 (48.0)	2811
Grey or Green	946 (1303)	1387 (1212)	746 (563)	53 (53.4)	3132
Brown	115 (357)	438 (332)	288 (154)	16 (14.6)	857
Total	2829	2632	1223	116	6800

$$\text{Hence } \chi^2 = \frac{(1768 - 1169)^2}{1169} + \frac{(807 - 1088)^2}{1088} + \ldots + \frac{(16 - 14.6)^2}{14.6}$$

$$= 1075.2 \text{ with } (4-1)(3-1) \text{ or 6 degrees of freedom.}$$

The chance of obtaining a value of χ^2_6 of 1075 or greater is extremely
small. It is in fact less than 1 in a million. Hence we must reject the
hypothesis that hair and eye colour are independent, and conclude
that there is some association.

11.4 Assume inoculation has no effect. The expected numbers on this assumption are shown in brackets.

	Not attacked	Attacked	Totals
Inoculated	276 (255.5)	3 (23.5)	279
Not inoculated	473 (493.5)	66 (45.5)	539
Totals	749	69	818

$$\chi^2 = \frac{(20.5)^2}{255.5} + \frac{(20.5)^2}{23.5} + \frac{(20.5)^2}{493.5} + \frac{(20.5)^2}{45.5}$$

$$= 29.6 \text{ with 1 degree of freedom.}$$

The probability of obtaining such a large value due to chance is extremely small. We therefore conclude that our assumption is wrong and that the inoculation is effective.

11.5 Here we have the result of 6 independent experiments, none of which is significant at the 5% level of significance. We may combine the tests by totalling the values of χ^2 and of the degrees of freedom. The result is $\chi^2 = 17.79$ with 6 degrees of freedom.
Values of χ^2_6 of up to 12.59 are expected at the 5% level of significance. The value obtained as a result of the six experiments is much greater than this. In fact, it exceeds the values which are expected at the 1% significance level. We therefore conclude that these results are unlikely to have occurred by chance. We therefore doubt the hypothesis.

11.6 Assume that this is a sample of 458 drawn at random from a population whose blood types are in the given proportions. The actual and expected frequencies are then as shown in the following table:

	M	MN	N	Total
Actual	375	77	6	458
Expected	380.14	74.20	3.66	458

$$\chi^2 = \frac{(5.14)^2}{380.14} + \frac{(2.80)^2}{74.20} + \frac{(2.34)^2}{3.66}$$

$$= 1.67 \text{ with 2 degrees of freedom.}$$

At the 5% level of significance with 2 degrees of freedom values of χ^2

of up to 5.99 are expected. Hence there is no reason to doubt our hypothesis.

There is no significant departure from the expected proportions.

11.7 Assume that laterality of hand and of eye are not connected. The expected values on this assumption are shown in brackets.

	Left-eyed	Ambiocular	Right-eyed	Totals
Left-handed	34 (35.43)	62 (58.55)	28 (30.02)	124
Ambidextrous	27 (21.43)	28 (35.41)	20 (18.16)	75
Right-handed	57 (61.14)	105 (101.04)	52 (51.82)	214
Totals	118	195	100	413

$$\chi^2 = \frac{(34 - 35.43)^2}{35.43} + \ldots + \frac{(52 - 51.82)^2}{51.82}$$

$$= 4.018 \text{ with } (3-1)(3-1) \text{ or 4 degrees of freedom.}$$

At the 5% level of significance with 4 degrees of freedom values of χ^2 up to 9.49 are expected. The results in the table could therefore easily have been obtained by chance on our assumption of independence. The given data is therefore not sufficient to establish any real association.

11.8 The table of actual numbers and the numbers expected on the assumption that there is no difference of opinion between the sexes is as follows:

	In favour	Against	Total
Women	216 (228.9)	84 (71.1)	300
Men	318 (305.1)	82 (94.9)	400
Total	534	166	700

$$\chi^2 = \frac{12.9^2}{228.9} + \frac{12.9^2}{305.1} + \frac{12.9^2}{71.1} + \frac{12.9^2}{94.9}$$

$$= 5.3$$

At the 5% level values of χ^2 with one degree of freedom up to 3.84 may be expected by chance. The value obtained exceeds this, although it does not exceed the value at the 1% level (6.63). We conclude that there probably is a difference of opinion between the sexes.

11.9 The table of actual numbers and the numbers expected on the assumption that there is no difference in the proportion following their father's occupation is as follows:

	Same occupation	Different occupation	Total
Doctors	34 (35.33)	166 (164.67)	200
Lawyers	19 (17.67)	81 (82.33)	100
Total	53	247	300

$$\chi^2 = \frac{1.33^2}{35.33} + \frac{1.33^2}{17.67} + \frac{1.33^2}{164.67} + \frac{1.33^2}{82.33}$$

$$= 0.17$$

A value of χ^2 with one degree of freedom as small as this could easily occur by chance. There is therefore no evidence here to establish a difference in the proportion applicable to each profession.

12.1 Assume that they are not. This then is a random sample from a population with median 23. Writing a $+$ sign for those above 23, a $-$ sign for those below and 0 for players aged 23 we obtain

$$+ - - - - - + +\ 0 - - - - -\ 0 - -$$

$$- - + - + - -\ 0 + - + - - + - +$$

Treating one of the 0s as $+$ and one as $-$ and neglecting the third we obtain

$$\text{Number of } + \text{ signs } = 10$$
$$\text{Number of } - \text{ signs } = 21$$
$$\text{Total } = 31$$

On our hypothesis, $+$ and $-$ signs are equally likely. The distribution of $-$ signs in samples of 31 has

$$\mu = 15.5$$
$$\sigma = \sqrt{31 \times \tfrac{1}{2} \times \tfrac{1}{2}} = 2.78$$

In our sample we have 21 $-$ signs, that is 5.5 above the mean. We therefore require the probability of obtaining 21 or more $-$ signs. Using normal distribution tables as an approximation and applying the continuity correction we obtain for the required probability the probability of obtaining a standard score exceeding $\frac{5.0}{2.78}$ or 1.8. This equals .0359. This result would occur by chance less than once in twenty times on our hypothesis. We must therefore suspect the hypothesis and conclude that players are getting younger.

12.2 Assume that there is no real difference between the two classes. On this hypothesis, we would expect class A to be superior in about one half of the cases and inferior in about one half.
From the data we find

Class A superior	11 students
Class A inferior	5 students
no difference	1 student — neglect.

On our hypothesis, in samples of 16, the number of superior students in class A has the distribution

$$\mu = 8$$
$$\sigma = \sqrt{16 \times \tfrac{1}{2} \times \tfrac{1}{2}} = 2$$

In this particular case we have 11 students i.e. 3 above the mean. Applying the continuity correction and using the normal tables as an approximation we obtain for the required probability the probability of a standard score exceeding $\dfrac{2.5}{2}$ or 1.25.

This result could reasonably have been obtained by chance and hence this test does not establish his claim. He has of course produced good results and hence criticism does not seem to be warranted.

Note: Being a statistics tutor, and finding that the sign test did not indicate a significant difference, he would doubtless have tried a paired *t* test if the marks could be assumed to be normally distributed. Again however he would not have obtained a significant result, and could not therefore establish his claim. His only alternative is to carry out the experiment once again. If he then repeated the performance i.e. if he obtained in both experiments, a total of 22 superior students, 10 inferior and 2 showing no difference then his claim might fairly be established as the probability of obtaining this overall performance or better purely by chance would be about 1 in 30.

12.3 Assume that there is no variation with age of mother in the sex ratio at birth.
For each age group we can then calculate the expected number of $+$ and $-$ signs (shown in brackets in the table) and compare them with the actual number.

Mother's Age	Number of		Total
	+ signs	— signs	
15 - 19	13 (10½)	8 (10½)	21
20 - 24	59 (52)	45 (52)	104
25 - 29	39 (45½)	52 (45½)	91
30 - 34	34 (39½)	45 (39½)	79
35 - 40	30 (35½)	41 (35½)	71

$$\chi^2 = 2\left[\frac{(2\frac{1}{2})^2}{10\frac{1}{2}} + \frac{7^2}{52} + \frac{(6\frac{1}{2})^2}{45\frac{1}{2}} + \frac{(5\frac{1}{2})^2}{39\frac{1}{2}} + \frac{(5\frac{1}{2})^2}{35\frac{1}{2}}\right]$$

= 8.17 with 5 degrees of freedom because the actual number of + signs in each of the 5 age groups is free to vary. At the 5% level of significance with 5 degrees of freedom values of χ^2 up to 11.07 are expected. Hence the result is not significant. A glance at the table shows that for ages under 25 the actual number of + signs exceeds the expected; at ages 25 and above the reverse is true. If we combine our data into these two groups we obtain:

Mother's Age	Number of		Total
	+ signs	— signs	
under 25	72 (62½)	53 (62½)	125
25 and over	103 (120½)	138 (120½)	241

Here $\chi^2 = 7.97$ for 2 degrees of freedom.

A departure of this magnitude or more from our hypothesis would occur about once in fifty times. It appears that masculinity of births is higher for young mothers.

12.4 The hypothesis to be tested is that these are four samples from the same population.
(i) Combine all the samples to form an artificial population and assume that the samples are drawn at random from this population.
(ii) The median of the four samples combined is 23.
(iii) Assigning a + to values above the median and a — to values below the median we obtain the following table of signs:

Age group	+	—	Total
1	3	7	10
2	4	6	10
3	5	5	10
4	8	2	10
Totals	20	20	40

(iv) The expected number of $+$ and $-$ signs in each age group are 5 and 5 on our hypothesis.
Hence

$$\chi^2 = \frac{(3-5)^2}{5} + \frac{(7-5)^2}{5} + \ldots + \frac{(2-5)^2}{5} = \frac{28}{5}$$

$$= 5.6 \text{ with 3 degrees of freedom.}$$

(v) Values of χ^2_3 up to 7.81 are expected at the 5% level.
(vi) We conclude that the samples could well be from the same population.

12.5 Assume exercise has no effect.
The measurements after exercise should then be greater or less than those before with roughly equal frequency. The results show

Greater after exercise   16

Less after exercise   4

On our hypothesis, the number greater after exercise has a distribution with

$$\mu = \quad 10$$

$$\sigma = \sqrt{20 \times \tfrac{1}{2} \times \tfrac{1}{2}} = 2.24$$

Our sample has a value of 16. This is 6 above the mean. Using normal tables as an approximation and applying the continuity correction we obtain for the probability of 16 or more the probability of a standard score greater than $\dfrac{5.5}{2.24}$ or 2.46. This probability (.0069) is less than 1 in 100 and hence we must suspect our hypothesis.
Exercise evidently affects the characteristic.

13.1 Here

$$\bar{x} = 6 \qquad \bar{y} = 5 \qquad \Sigma(x - \bar{x})(y - \bar{y}) = 40$$

$$s_x = 2\sqrt{2} \quad s_y = 2\sqrt{2}$$

Hence

$$r = \frac{40}{5 \times 2\sqrt{2} \times 2\sqrt{2}} = 1$$

This is a case of perfect correlation, which is obvious from the given figures.

13.2 Here

$$\bar{x} = 6 \qquad \bar{y} = 5 \qquad \Sigma(x - \bar{x})(y - \bar{y}) = -40$$

$$s_x = 2\sqrt{2} \quad s_y = 2\sqrt{2}$$

Hence

$$r = \frac{-40}{5 \times 2\sqrt{2} \times 2\sqrt{2}} = -1$$

This is a case of perfect negative correlation, which is obvious from the given figures.

13.3 Here $\quad \bar{x} = 3 \qquad \bar{y} = 4 \quad \Sigma(x - \bar{x})(y - \bar{y}) = 0$

$$s_x = \sqrt{2} \qquad s_y = 2$$

Hence $\quad r = 0.$

Therefore x and y are uncorrelated as is obvious from the figures since y takes the same value irrespective of whether $x = 1, 2, 4$ or 5.

13.4 Here $\Sigma x = 122 \quad \Sigma y = 123 \qquad \Sigma xy = 806$

$$\Sigma x^2 = 832 \quad \Sigma y^2 = 843 \quad \frac{1}{n}\Sigma xy = 42.42$$

$$\bar{x} = 6.421 \qquad \bar{y} = 6.474$$

$$s_x = 1.60 \qquad s_y = 1.566$$

$$\therefore r = \frac{42.42 - 6.421 \times 6.474}{1.60 \times 1.566} = 0.34$$

The analysis proceeds as follows:

(*i*) Assume x and y are uncorrelated (i.e. $\rho = 0$) and are normally distributed.

(*ii*) The sample value of $r = 0.34$, and the corresponding value of w (from the tables) $= 0.354$.

(*iii*) On our hypothesis, the value of w should be normally distributed with mean 0 and SD $\dfrac{1}{\sqrt{16}} = 0.25$

(*iv*) This sample therefore has a value of w which is $\dfrac{0.354}{0.25}$ or 1.42

standard deviations above the mean. A correlation of this amount or greater is likely to occur nearly 8 times in 100 due to chance alone.

This sample does not, therefore, establish any correlation between english and history marks. This is also suggested by the scatter diagram.

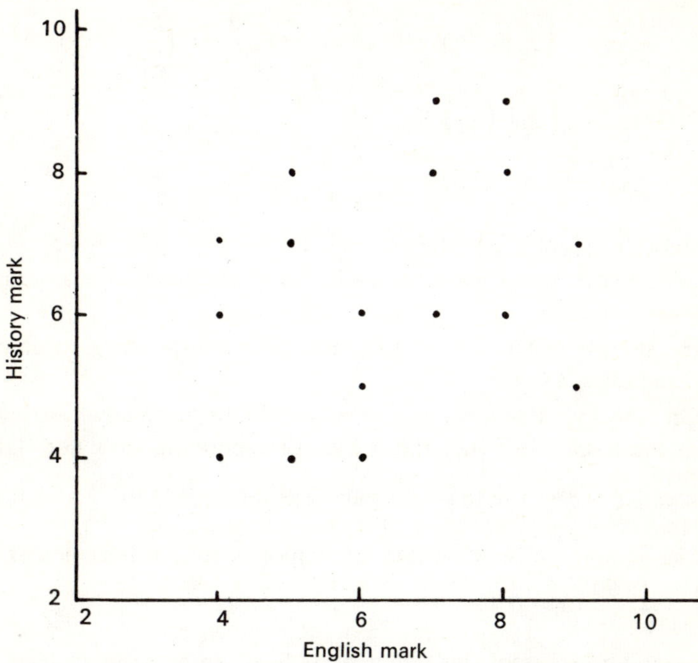

Scatter diagram for the english and history marks of 19 students

13.5

Age of bridegrooms (y)	Age of brides (x)						b	$f_y b$	$f_y b^2$	$b\Sigma fa$
	—20	20—4	25—9	30—4	35—	Total (f_y)				
under 20	4					4	—1	—4	4	4
20 — 24	14	21	1			36	0	0	0	0
25 — 29	3	12	4	1		20	1	20	20	3
30 — 34	1	3	2	1	1	8	2	16	32	12
35 —		1	1	2	7	11	3	33	99	78
Total (f_x)	22	37	8	4	8	79		65	155	97
a	—1	0	1	2	3					
$f_x a$	—22	0	8	8	24	18				
$f_x a^2$	22	0	8	16	72	118				
$a\Sigma fb$	—1	0	11	18	69	97				

$$s_a{}^2 = \frac{118}{79} - \left(\frac{18}{79}\right)^2 = 1.442 \qquad s_b{}^2 = \frac{155}{79} - \left(\frac{65}{79}\right)^2 = 1.285$$

$$r = \frac{\dfrac{97}{79} - \left(\dfrac{18}{79}\right)\left(\dfrac{65}{79}\right)}{\sqrt{1.442} \times \sqrt{1.285}} = +0.76$$

The analysis proceeds as follows:

(*i*) Assume that x and y are normally distributed and have a value of ρ of 0.8.

(*ii*) The sample value of $r = 0.76$, and the corresponding value of w from the tables is 0.996.

(*iii*) On our hypothesis values of w should be normally distributed with a mean of 1.099 (i.e. the value corresponding to a correlation coefficient of 0.8 in the table on page 259) and a SD of $\dfrac{1}{\sqrt{76}}$ or 0.115.

(*iv*) The sample value of w thus corresponds to a standard score of $\dfrac{0.996 - 1.099}{0.115}$ or -0.90.

(*v*) A standard score of this size is quite likely to occur by chance. We therefore have no reason on this information to dispute the hypothesis that the correlation between ages of brides and bridegrooms is 0.8.

13.6a. Write x for the number of marriages in thousands and y for the production of brown coal in millions of tons.

Then
$$\Sigma x = 888.0 \qquad \Sigma y = 192.9 \qquad n = 10$$
$$\Sigma x^2 = 80166.64 \quad \Sigma y^2 = 3814.35 \quad \Sigma xy = 17463.16$$

$$s_x = \sqrt{\frac{80166.64}{10} - 88.8^2} = 11.46$$

$$s_y = \sqrt{\frac{3814.35}{10} - 19.29^2} = 3.055$$

$$\therefore r = \frac{\dfrac{17463.16}{10} - 88.8 \times 19.29}{11.46 \times 3.055} = 0.95$$

b. This value of r corresponds to a value of w of 1.832. If $\rho = 0$ values of w should be nearly normally distributed about a mean of zero and SD of $\dfrac{1}{\sqrt{7}}$ or 0.378. Hence the value for this sample corresponds to a standard score of $\dfrac{1.832}{0.378}$ or 4.85. This is most unlikely to have occurred by chance and we conclude that there is correlation between x and y.

This however is almost certainly spurious correlation and not the result of cause and effect. It illustrates the difference between correlation and causation.

14.1 The differences are

$$1 \quad -1 \quad 0 \quad 1 \quad -1 \quad 1 \quad 1 \quad 0 \quad -1 \quad -1$$

$\Sigma d \quad = 0$ as it should.

$\Sigma d^2 \quad = 8$

$$\therefore r_s \;=\; 1 - \frac{6}{990} \times 8 = 0.952$$

The probability of a value of .746 or more due to chance alone is 0.01. Hence it is extremely unlikely that this result (0.952) would have been obtained by chance.

14.2 The differences are

$$-3 \quad 3 \quad 3 \quad -2 \quad -3 \quad -3 \quad 0 \quad 4 \quad -2 \quad 0 \quad 1 \quad -1 \quad 1 \quad 1 \quad 1$$

$\Sigma d \quad = 0$ as it should

$\Sigma d^2 \quad = 74$

$$\therefore r_s \;=\; 1 - \frac{6}{3360} \times 74 = 0.868$$

The probability of a value of .623 or more due to chance is 1 in 100. Hence this result is highly significant and suggests that examinations are not really necessary. It is possible that other tutors may not have the same judgement and it would therefore be advisable to test estimates from other tutors before abandoning the examination system.

Also the examination system allows comparisons to be made between *all* students. If examinations were abandoned it would be difficult to rate against one another candidates assessed by different tutors.

14.3 What we require is some measure of the difference between each tutor's placings and the examination placings. Σd^2 is such a measure. All we need to do is to calculate Σd^2 for each tutor and rank the values of Σd^2 in order from the smallest to the largest. These values place the tutors in order of their success in estimating the students' examination performances.

14.4 The relative ranks (allowing for equal rankings) are

Indispensability	1	2	3	4	5	6	7	8	9	10	11	12	13	14	15
Salary	1	$2\frac{1}{2}$	$2\frac{1}{2}$	10	15	12	10	$4\frac{1}{2}$	6	8	$13\frac{1}{2}$	10	$13\frac{1}{2}$	$4\frac{1}{2}$	7
The differences are:	0	$-\frac{1}{2}$	$\frac{1}{2}$	-6	-10	-6	-3	$3\frac{1}{2}$	3	2	$-2\frac{1}{2}$	2	$-\frac{1}{2}$	$9\frac{1}{2}$	8

$$\Sigma d \quad = 0 \text{ as it should.}$$

$$\Sigma d^2 \quad = 371.5$$

$$\therefore r_s \quad = 1 - \frac{6}{3360} \times 371.5 = 0.337$$

There is a positive correlation between the rankings but not a significantly high one.

A study of correlation between these two variables is of no real practical value. There could be very good reasons why an employee on a low salaried job is indispensable to a firm. Conversely there are situations where a highly paid position can fairly easily be dispensed with, e.g. a sales manager in wartime when goods are rationed, or where the three senior positions can be carried on by two persons with appropriate delegation.

The test is hardly appropriate, although we would expect some positive correlation in practice.

15.1a. For this sample

$$\overline{x} = \frac{588}{12} = 49 \text{ cm}$$

Our estimate of the population mean is therefore 49 cm.

b. & c. Heights jumped may be assumed to be nearly normally distributed and hence we may use the t distribution.

$$s = \sqrt{\frac{29234}{12} - 49^2} = 5.93 \text{ cm}$$

giving

$$t = \frac{\overline{x} - \mu}{5.93} \sqrt{11}$$

We see from t distribution tables that where there are 11 degrees of freedom, 95% of samples have a value of t numerically less than 2.201. Hence 95% of samples will have a value of $\bar{x} - \mu$ numerically less than $\dfrac{2.201}{\sqrt{11}} \times 5.93$ or 3.94. That is, the mean of 95% of samples lies within 3.94 cm of μ.

Therefore on the basis of the mean of our sample, namely 49 cm , we are 95% confident that the value of μ lies between 45 and 53 cm. The 95% confidence limits for the population mean based on this sample are therefore 45 and 53 cm. In answer to $b.$ we may therefore state that in 19 samples out of 20 the error would be less than 4 cm.

15.2 With a sample of this size we may assume that $\bar{x}$ is normally distributed with mean μ and we may estimate the standard deviation of $\bar{x}$ as

$$\frac{s}{\sqrt{n}} \quad \text{or} \quad \frac{5}{\sqrt{100}} \quad \text{or } 0.5$$

Hence 95% of samples will have a mean which lies within 1.96×0.5 or 0.98 cm of μ. Our sample has a mean of 48 cm. Hence we are 95% confident that the population mean lies between 47 and 49 cm.

Hence the answers are

$a.$ 48 cm

$b.$ In 19 samples out of 20 the error would be less than 1 cm

$c.$ 47 and 49 cm.

$15.3a.$ For this sized sample the proportion in the sample will be normally distributed with mean π and standard deviation $\sqrt{\dfrac{\pi(1-\pi)}{400}}$

where π is the population proportion.

Hence in 95% of samples the proportion will differ from π by less than $1.96 \sqrt{\dfrac{\pi(1-\pi)}{400}}$

Ours may be the one unusual sample in 20 but we are 95% confident that the error in using $\dfrac{256}{400}$ or 0.64 as an estimate of π is less than $1.96 \sqrt{\dfrac{\pi(1-\pi)}{400}}.$

We do not know the value of π to substitute in this formula but with large samples it is near enough to use the sample proportion. This gives

$$1.96 \sqrt{\frac{0.64 \times 0.36}{400}} \text{ or } 0.047$$

Hence the 95% confidence limits for π, based on our sample, are 0.59 and 0.69.

b. Let n be the sample size.

Again if we use 0.64 as an approximation for the value of π in the formula for the standard deviation we obtain

$$1.96 \sqrt{\frac{0.64 \times 0.36}{n}} = 0.02$$

$$\therefore n = \frac{1.96 \times 1.96 \times 0.64 \times 0.36}{0.02 \times 0.02} = 2213$$

A sample of about 2,200 would be required.

3:   Answers to Miscellaneous Problems

Note:P is used as an abbreviation for the word 'probability'.

1a. mean   = 6 metres   *b.* mean   = 600 cm
   median = 6 metres   median = 600 cm
   SD   = 2 metres   SD   = 200 cm

2.a. 0.0228   *b.* 0.4967   *c.* 0.0548   *d.* 42

3.a. 334   *b.* 2398

4. The observed number is 66 SDs above the mean. This is a most unlikely event. Has the machine become faulty?

5.a. Mark   3 4 5 6 7 8 9
      Frequency   1 2 5 5 3 3 1
   b. Mean $= 6$;   SD $= 1.52$

6. 4902 to 5098 is one of the many possible solutions.

7.a. 0.05   *b.* 0.45   *c.* 0.95   *d.* 0.95

8.a. —0.2, +1.0, +3.0, —0.8   *b.* 0.24, 0.50, 0.07, 0.33
   c. N(60, 5)   *d.* $1\frac{1}{2}$ SDs from the mean. It could
                                         easily occur by chance.

9.a. 0.9332, 0.6179   *b.* 0.871, 0.9955

10.a. 0.9772   *b.* 0.0049

11.a. Yes. Green is higher 3 times out of 10. This is not inconsistent with the hypothesis.
   b. $r_s = .939$ which is too high reasonably to have happened by chance.

12. No. On the null hypothesis the probability of this difference $= 0.003$.

13.a. 0.0018.

   b. Yes. On the null hypothesis the probability of this difference or more is much less than 1 in 1000.

14.a. 0.0062

   b. Professional rallies are significantly shorter.

15. Yes. $x_1^2 = 7.49$ for which $P < 0.01$.

16.a. Sign test gives no such evidence. *t* test gives $t_{14} = 2.29$ or $P = 0.02$. Player 1 seems to be a significantly better player.

   b. On the hypothesis that mean $= 500$, $t_{14} = 2.27$. It seems unlikely that these players are from a population with mean 500. Certainly for them 'most games' do not reach 500. Did you notice that they seem to become less 'defensive' in the last few games?

17.a. $29\frac{1}{2}$ *b.* Yes. $\chi_2^2 = 3.2$ which is not significant.

18. $29\frac{1}{2}$; $23\frac{1}{2}$; $35\frac{1}{2}$; $38\frac{1}{2}$.

19. The estimated chances must be incorrect. $x_6^2 = 54$. This is highly significant.

20.a. 2.18 a.m.   *b.* Yes. The difference is highly significant.

21. For a two tailed test $P = 0.087$. It could easily happen by chance.

22.a. $\dfrac{1485}{3000}$ *b.* $\dfrac{770}{1576}$ *c.* $\dfrac{683}{1515}$ *d.* $\dfrac{363}{628}$

    e. $\dfrac{1424}{3000}$ *f.* $\dfrac{1620}{3000}$ *g.* $\dfrac{1200}{3000}$ *h.* $\dfrac{459}{866}$

    i. $\dfrac{362}{802}$ *j.* $\dfrac{207}{363}$

23. Yes. $r_s = .709$, nearing the .01 significance level.

24. For this sample $\chi_{99}^2 = 144$ which is significant at the 1% level. It is thus not consistent with the hypothesis.

25. On the employee's hypothesis this sample gives $t_9 = 1.33$ which is consistent with the hypothesis.

26. Yes. The chance of this happening with the advertised failure rate is extremely small.

27a. The chance of a difference between sample and population mean of this amount or greater $= 0.06$. We are not justified in concluding that these female university students are definitely more garrulous than people generally.

b. No. The t test gives $t_{24} = 0.57$ on the null hypothesis, which is not significant at the 5% level.

28. Yes. Null hypothesis gives $P = 0.003$.

29.a. If performance length is normally distributed, a performance of at least this length would occur 7 times in 100.

b. It reinforces the belief that the conductor uses a slow tempo. Here $P = 0.001$.

30.a. About 55 times in 1000.

b. About 5 seconds after 7.43 a.m.

31. Yes. On the null hypothesis P is very small.

32. Yes. The difference between the proportions is highly significant $(P < 0.001)$.

33. Yes. $t_{16} = .28$. Even if the distribution is far from normal this does not indicate a significant difference.

34.a. Day students: No significant difference $(\chi_4^2 = 4.49)$

      Evening students: Highly significant difference $(\chi_4^2 = 98)$

  b. University A $=$ Highly significant difference $(\chi_2^2 = 269)$

      University B $=$ Highly significant difference $(\chi_2^2 = 177)$

      University C $=$ Highly significant difference $(\chi_2^2 = 69)$

36. Yes. The difference in the proportions is significant $(P = 0.006)$.

37. Yes. The difference equals 3.75 SDs.

38. Yes. The difference equals 21 SDs.

39.a. $\dfrac{1}{1024}, \dfrac{1023}{1024}, \dfrac{1}{4}, \dfrac{1}{4}.$

  b. 370.4.

40. Yes. χ_2^2 for independence is extremely high. There is a strong tendency for females to enter the arts faculty.

41. There is a significant increase in the height ($t_9 = 4.6$). We cannot tell whether this will improve its market value.

42.a. 0.9896   *b.* At intervals of 740 hours.

43. $\bar{x} = 5.368$, $s = 2.612$, mean deviation $= 2.206$

44.a. $r = -0.0075$
   b. We do not know whether the variables are normally distributed.
   c. $r_s = -0.104$ which is not significant.

45. Testing the mean gives $t_{64} = 4.95$ which is highly significant. Testing the SD gives $\chi^2_{64} = 45.9$ which is not significant. Evidently the mean length has altered but there is no evidence of a significant change in the standard deviation.

46.a. $(\frac{1}{2})^{14}$
   b.(i) The differences between individual scores yield $t_{31} = 0.33$ which is not significant.
   (ii) The difference between means yields $t_{62} = 3.41$ which is significant ($P < 0.001$). This depends on the population being near normal. A median test yields $\chi^2_1 = 10.4$. This is again significant ($P < 0.001$).
   (iii) There is no significant correlation ($r = -0.075$).
   (iv) In period one all scores are on or below the median. In period 8 all scores except one are above the median. This does indicate a significant difference but this is due to the sex difference established in *(ii)*.
   (v) The difference between mean scores in the two halves $= 13.219 - 11.156 = 2.063$. Estimating the variance of the population of scores from the sample of 64 gives 1.8 for the SD of the difference of means of the two halves. Hence the difference is not significant. (*Note* a normal assumption is not required in view of the size of the combined sample.)
   (vi) The differences between individual scores yield $t_{15} = 1.52$ for girls and 0.29 for men. Neither is significant.
   (vii) Ignoring zeros there are 15 plus signs and 9 minus signs. This could easily occur due to chance. Hence the tendency mentioned is not established.

Appendix II

1: Logarithms

1. Introduction

In statistics (and in many other subjects) we need to know how to multiply and divide. There are many ways of doing this.

(*1*) Long multiplication and division — as at school

(*2*) With a slide rule (approximate)

(*3*) Using logarithm tables

(*4*) Using calculating machines.

Method (*1*) is always possible and will enable the student to obtain the answers to problems in this book. Methods (*2*), (*3*) and (*4*) are generally quicker and the student should try to become familiar with them because the ultimate saving in time is great.

2. The value of logarithm tables

Most people find addition and subtraction much easier and much quicker than multiplication and division. The use of logarithms is simply *a device for substituting addition and subtraction for multiplication and division* and hence it speeds up those processes.

3. The tables

For every positive number there is a logarithm and for every logarithm there corresponds a positive number — just as for every person there is a fingerprint and for every fingerprint there corresponds a person. To use the fingerprints system we need a list of names in alphabetical order with the fingerprints opposite and also a list of fingerprints in some logical order with the name opposite. Likewise to use logarithms we need:

(*1*) a table with numbers in order and their logarithms opposite — called 'logarithm tables', and

(*2*) a table with logarithms in order and the corresponding numbers opposite — called 'antilogarithm tables'.

We can then find for any number, its logarithm, or for any logarithm, its number.

4. Logarithm

The logarithm of any number consists of two parts:
(*1*) a *positive* decimal part (i.e. less than unity) which corresponds with the actual figures that make up that number (it is for this only we need the tables), and
(*2*) a *positive or negative* integer which tells us whère amongst the figures that make up the number the decimal point lies. *Note:*This positive or negative integer is found by subtracting one from the number of figures to the left of the decimal point in the number.

5. Given a number — to find its logarithm

If we look up the number 3463 in logarithm tables (written log tables for short) we find corresponding to it 5395. This means the log of 3463 is

$$3 + .5395.$$

The 3 is there simply to indicate where the decimal point lies in the number.

Here are some numbers and their logarithms —

Number	Logarithm
3463	3 + .5395
346.3	2 + .5395
34.63	1 + .5395
3.463	0 + .5395
.3463	—1 + .5395
.03463	—2 + .5395
and so on	
151.1	2 + .1793
26.4	1 + .4216
.2562	—1 + .4085
.00821	—3 + .9143

6. Given the logarithm — to find the number

Here we use *the antilog tables.*
If the given logarithm is 2 + .5395 we look up .5395 in the antilog tables and find the figures 3463 opposite. The 2 tells us there should be three figures to the left of the decimal point. Hence 346.3 is the number

whose logarithm is 2 + .5395. Similarly —

Logarithm	Number
3 + .5395	3463
2 + .5395	346.3
1 + .5395	34.63
0 + .5395	3.463
—1 + .5395	.3463
—2 + .5395	.03463
2 + .1793	151.1
1 + .4216	26.4
—1 + .4085	.2562
—3 + .9143	.00821

7. *How to multiply using logarithm tables*

If two numbers are *multiplied* together the log of the product is the *sum* of the logarithms of the two factors. (It is this curious property that enables us to substitute the simpler process of addition for that of multiplication.) You can verify this from the tables. Thus

$$\log 2 = 0 + .3010$$
$$\log 4 = 0 + .6021$$

But $(0 + .3010) + (0 + .6021)$

$$= 0 + .9031$$
$$= \log 8$$
$$= \log 2 \times 4$$

Thus $\log 2 \times 4 = \log 2 + \log 4$

Hence the rule: To multiply, add the logarithms.

Examples

a. Multiply 11 by 12.

Number		Logarithm
11	$\longrightarrow$	1 + .0414
12	$\longrightarrow$	1 + .0792
132.0	$\longleftarrow$	2 + .1206 adding

Therefore $11 \times 12 = 132$.

$\longrightarrow$ indicates that from the number the log is obtained

(i.e. log tables)

⟵——— indicates that from the log the number is obtained

(i.e. *antilog* tables)

b. Multiply 127 by 62.

	Number		Logarithm
	127	⟶	2 + .1038
	62	⟶	1 + .7924
	7874	⟵	3 + .8962 adding

Therefore $127 \times 62 = 7874$

c. Multiply 824 by 75

	Number		Logarithm
	824	⟶	2 + .9159
	75	⟶	1 + .8751
			3 + 1.7910 adding
	61800	⟵ =	4 + .7910

Therefore $824 \times 75 = 61800$.

d. Multiply 62.5 by .234 by .056

	Number		Logarithm
	62.5	⟶	1 + .7959
	.234	⟶	−1 + .3692
	.056	⟶	−2 + .7482
			−2 + 1.9133 adding
	.8191	⟵ =	−1 + .9133

Therefore $62.5 \times .234 \times .056 = .8191$.

8. *How to divide using logarithms*

Here the rule is: To divide, subtract the logarithms.

We thus substitute the simpler process of subtraction for division. Again from the tables we can check that this holds.

Thus
$$\log 8 = 0 + .9031$$
$$\log 4 = 0 + .6021$$

Subtracting $(0 + .9031) - (0 + .6021) = 0 + .3010$
$$= \log 2$$
$$= \log \frac{8}{4}$$

Hence
$$\log \frac{8}{4} = \log 8 - \log 4$$

Thus to find the result of dividing two numbers, we subtract the log of the denominator from the log of the numerator and look up this difference in antilog tables to find the result of the division.

Examples

a. Divide 240 by 3

	Number		Logarithm
	240	$\longrightarrow$	$2 +$ .3802
	3	$\longrightarrow$	$0 +$ .4771
			$2 -$ .0969 subtracting
	80.00	$\longleftarrow$ $=$	$1 +$ .9031

Therefore 240 divided by 3 $= 80$.

(*Note:* We need to express $2 - .0969$ in the form of an integer plus a *positive* decimal in order to use the antilog tables. We therefore write $2 - .0969$ in the equivalent form $1 + .9031$.)

b. Divide 127 by 142.

	Number		Logarithm
	127	$\longrightarrow$	$2 +$ .1038
	142	$\longrightarrow$	$2 +$ .1523
			$0 -$ .0485 subtracting
	.8943	$\longleftarrow$ $=$	$-1 +$ .9515

Therefore 127 divided by 142 $= .8943$.

9. How to find square roots using logarithms

It can be shown that the logarithm of the square root of a number equals one half of the logarithm of that number. This fact may be used to calculate square roots of numbers.

Examples

a. Find $\sqrt{5}$.

	Number		Logarithm
	$\sqrt{5}$	$\longrightarrow$	$\frac{1}{2}(0 + .6990)$
	2.236	$\longleftarrow$ $=$	$0 + .3495$

Therefore $\sqrt{5} = 2.236$.

b. Find $\sqrt{17}$.

Number		Logarithm
$\sqrt{17}$	$\longrightarrow$	$\frac{1}{2}(1 + .2304)$
		$= \quad .5 + .1152$
4.123	$\longleftarrow$	$= \quad 0 + .6152$

Therefore $\sqrt{17} = 4.123$.

2:   Coin Tossing Experiment

The following summary of the results of a coin tossing experiment conducted by students in 1967 is included here because it illustrates a number of the topics discussed in the book. Each student attending tutorials in a particular week was asked to toss a coin once, and record whether a tail was obtained, then to toss it twice, recording the number of tails, then three times, four times, five times, ten times and finally one hundred times. The results of these experiments were collected from the students and analysed.

Each toss of an unbiased coin can result in either a head or a tail, each of these outcomes having probability $\frac{1}{2}$ or 0.5. When a coin is tossed n independent times, therefore, the number of tails obtained should follow a binomial distribution with the value of π equal to 0.5. It is therefore possible to test whether a particular set of observed tosses is reasonably in accordance with the hypothesis that $\pi = 0.5$. A suitable test is the χ^2 test, in which the observed frequencies are compared with those expected under the hypothesis. The results are shown in the Table. The expected frequencies are calculated from the theoretical probabilities. Thus, in the case where three coins are tossed, it can easily be seen that the probability of obtaining no tails is $\frac{1}{8}$, as is the probability of obtaining no heads (i.e. 3 tails); and the probabilities of getting one tail or two tails are each $\frac{3}{8}$. Hence, in this case, since there were altogether 262 people who submitted results for the three-coin toss, we would expect $\frac{1}{8} \times 262 = 32.75$ (or 32.8 rounded to one decimal place) occurrences of no tails, $\frac{3}{8} \times 262 = 98.25$ occurrences of exactly one tail, 98.25 occurrences of exactly two tails and 32.75 occurrences of exactly three tails. The value of χ^2 obtained for the particular results observed (see below) is then

$$\frac{(25 - 32.75)^2}{32.75} + \frac{(105 - 98.25)^2}{98.25} + \frac{(101 - 98.25)^2}{98.25} + \frac{(31 - 32.75)^2}{32.75}$$

or 3.39, which should be a value of χ^2 with $4-1=3$ degrees of freedom. In fact this is not an unlikely value for such a χ^2, and so we are prepared to accept the hypothesis that $\pi = \frac{1}{2}$ or in other words that there was no bias in the results.

Comparison of observed with expected frequencies in a coin tossing experiment

	Number of tails	Observed frequency	Expected frequency	x^2	Degrees of freedom
one coin tossed	0 1	135 130	132.5 132.5	0.094	1
two coins tossed	0 1 2	48 137 59	61 122 61	4.68	2
three coins tossed	0 1 2 3	25 105 101 31	32.8 98.2 98.2 32.8	3.39	3
four coins tossed	0 1 2 3 4	15 63 88 58 18	15.1 60.5 90.8 60.5 15.1	0.851	4
five coins tossed	0 1 2 3 4 5	4 27 94 89 48 6	8.4 41.8 83.8 83.8 41.8 8.4	9.95	5
ten coins tossed	0 1 2 3 4 5 6 7 8 9 10	0 2 5 27 53 68 55 36 12 1 0	0.3 2.5 11.4 30.4 53.1 63.7 53.1 30.4 11.4 2.5 3	7.00	10

It can easily be verified from the tables that the respective values of χ^2 are all 'reasonable'. That is, in each case there is a chance greater than 5% of observing a higher value of χ^2 merely by random variation.

(In other words, none of the observed values of χ^2 is 'significant at 5% level'.) It would therefore be reasonable to conclude that there was no significant tendency in the class as a whole towards the use of biased coins. Note that this does *not* imply that individual students did not corrupt their results, but merely states that as a whole there was no significant trend.

A comparison of the observed means and standard deviations (of the number of tails) with those of the corresponding populations may also be of interest.

Number of coins tossed	Observed mean	Population mean	Observed SD	Population SD
1	0.491	0.5	0.4999	0.5000
2	1.045	1.0	0.6608	0.7071
3	1.527	1.5	0.8029	0.8606
4	2.004	2.0	1.0227	1.0000
5	2.627	2.5	1.0034	1.1181
10	5.143	5.0	1.4644	1.5811

The standard deviations are, with one exception, smaller than the theoretical population values. This tends to confirm the impression gained from inspection of the results that there was a tendency for students not to report outlying observations. Further evidence for this is given by the fact that in all but one case, the reported frequencies of the extreme values are less than those expected. A sign test could have been used to test this aspect.

The results of the 100 toss case are as follows:

No. of tails	Frequency
37	2
38	2
39	1
40	2
41	3
42	10
43	3
44	13
45	10
46	18
47	20
48	22
49	24
50	13
51	22
52	23
53	18
54	12
55	14
56	11
57	10
58	8
59	4
60	2
61	2
62	2
63	0
64	1
65	0
66	1
67	0
68	1
69	1

The mean of these observed results is 50.276, and the standard deviation 5.142.

The corresponding population figures are of course 50 and

$$\sqrt{100 \times \tfrac{1}{2} \times \tfrac{1}{2}} = 5.$$

It is of interest to note the proportions lying within 1 and 2 standard deviations from the observed mean. The proportion lying within 1 standard deviation (i.e. lying between 46 and 55 inclusive) is

$$\frac{186}{275} = 0.6763,$$

which is very close to the expected proportion, found from the normal tables, of $2 \times 0.3413 = 0.6826$. Similarly for 2 standard deviations, the observed proportion is

$$\frac{262}{275} = 0.9527,$$

compared with the population proportion of 0.9546. The closeness of the observed proportions to the population proportions is obviously very satisfactory.

Students who wish to obtain a more comprehensive assessment of these results could well perform a χ^2 test for themselves comparing the reported frequencies with the corresponding expected frequencies. (*Note* It is unwise to perform a χ^2 test on data where the expected frequency of a class is less than 5. It is, however, reasonable to group together several such classes so that the expected frequency of the new class so formed is greater than 5.)

The results were collected from 18 of the 20 tutorial groups, and the mean number of tails within each of these groups was as follows.

Group number	Mean number of tails
2	51.333
3	48.500
4	50.000
5	49.714
6	52.071
7	49.700
8	52.917
9	49.500
10	51.667
11	50.063
12	49.421
13	50.333
14	49.750
15	50.056
17	48.429
18	49.842
19	50.667
20	50.625

It can be seen that these results are all close to the theoretical population mean of 50, and in fact the standard deviation of these results is 1.118. This illustrates the important fact that the variability of a sample *mean* is much less than that of a single observation. One primitive indication of variability, for instance, is the range. The range of the sample means is 48.429 to 52.917, i.e. 4.5 roughly, compared with 37 to 69, i.e. 32, for the individual counts. In fact, if the population standard deviation is σ then the standard deviation of the 'population of means of samples of size n from the original population' is $\dfrac{\sigma}{\sqrt{n}}$. In this case, there were about 16 people in each group (the largest group was 24, and the smallest 8) and so the standard deviation of the observed means of the groups could have been predicted to be about $\dfrac{5}{\sqrt{16}} = 1.25$, which is quite close to the observed value of 1.118.

Conclusions

The results obtained have been in accordance with the theory. The most important aspects are these:

1. In repeated sampling, the characteristics of the sample resemble closely the corresponding characteristics of the population.

2. In particular, the observed frequencies of occurrence of particular values are close to those expected, and the mean and standard deviation of the sample are close to the respective population parameters.

3. The means of a set of samples are less variable than the individual items in the samples; in fact, if the standard deviation of the population is σ, the standard deviation of means of samples of size n from the population will be $\dfrac{\sigma}{\sqrt{n}}$, and so the reliability of the mean of a sample is proportional to the square root of the size of sample.

3: Test Papers

Test papers **A1** to **A6** do not require a knowledge of the text beyond Chapter 8. About one hour should be sufficient for each test.

Test Papers **B1** to **B6** require a knowledge of the whole text. About two hours should be sufficient for each of these tests.

A1

1. For the following data calculate:

a. The arithmetic mean

b. The median

c. The mode

d. The upper and lower quartiles

e. The semi-interquartile range

f. The mean deviation

g. The standard deviation.

Employee identification number	Number of days absent
001	5
002	0
003	1
004	7
005	1
006	2
007	9
008	5
009	1
010	3

2. The marks obtained by students in a test were as follows:

Marks	Number of students
10–14	0
15–19	2
20–24	10
25–29	9
30–34	6
35–39	3

Determine *a.* The mean mark
       b. The standard deviation
       c. The 70th percentile.

3. The proportion of families in a community with various numbers of children are as follows:

Number of children	0	1	2	3	4	5 or more
Proportion	0.10	0.10	0.20	0.25	0.20	0.15

a. What is the probability that a family selected at random will have either no children or 4 or more?

b. What is the chance that two families selected at random will both have exactly two children?

4. An aptitude test for graduate study in a certain field yields results that approximate the normal distribution with mean 500 and standard deviation 100.

a. What percentage of test scores would be located between 300 and 600?

b. Of 10,000 test scores, how many would we expect to be

    (*i*)   225 or lower?
    (*ii*)   675 or greater?

c. Suppose the top 10 per cent of the scores are to be regarded as distinctly superior. What is the lowest score that belongs to this category?

d. If we were to test students in random groups of 16 and if as a matter of interest we recorded the mean test score of each group, how would this mean be distributed?

A2

1.a. Calculate the median, mean and standard deviation of the numbers

$$1, \quad 3, \quad 6, \quad 10, \quad 15.$$

b. The numbers 10, 30, 60, 100 and 150 are ten times the numbers in *a.* Hence write down the mean and standard deviation of these larger numbers.

c. By adding 51 to the numbers in *a.* we obtain 52, 54, 57, 61 and 66. Hence write down the mean and standard deviation of these numbers.

2. In a class of 48 the examination marks ranged from 45 to 70 with the following frequency distribution:

Mark	Number of students with this mark
45	2
50	3
55	20
60	16
65	6
70	1
Total	48

Convert these marks to standard scores (i.e. number of standard deviations from the mean).

3. The Intelligence Quotient (IQ) of the individuals in a population is normally distributed with a mean of 100 and a standard deviation of 20.

a. What proportion of the population will have an IQ greater than 130?

b. In a school of 400 students, how many students would you expect to have an IQ between 90 and 120?

c. What is the chance that two individuals selected at random will both have an IQ exceeding 130?

d. What is the standard deviation of the means of the IQs of groups of 100 individuals?

4. In a spelling test 30 students were asked to spell a large number of words. For convenience of recording, a maximum possible mark of 10 was selected. Persons obtaining from $\frac{1}{2}$ to less than $1\frac{1}{2}$ were given the mark 1, those obtaining from $1\frac{1}{2}$ to less than $2\frac{1}{2}$ were given the mark 2 and so on.

The marks thus obtained by the 30 students were

$$6, 1, 5, 2, 7, 3, 5, 5, 8, 6, 4, 9, 6, 6, 3,$$
$$8, 7, 5, 6, 4, 2, 8, 4, 9, 7, 3, 5, 6, 1, 7.$$

a. Record the marks in the form of a frequency table.
b. Calculate the mean from the frequency table.
c. Give (without simplification) a numerical expression for the standard deviation.
d. Draw a less than ogive.
e. Calculate from the ogive the median, the upper and lower quartiles, the ninety percentile.

A3

1. A company selling rope finds that during the day the following lengths of rope were sold — measured in metres:

$$4, 6, 5, 3, 9, 8, 9, 4, 6, 6.$$

a. Calculate the mean, the median and the standard deviation of these lengths.
b. If these lengths had been measured in centimetres they would, of course, have been:

$$400, 600, 500, 300, 900, 800, 900, 400, 600, 600$$

What are the mean, the median and the standard deviation of the lengths measured in centimetres?

2. The marks obtained in a class examination were:

$$5, 4, 5, 7, 6, 8, 3, 6, 8, 6,$$
$$4, 5, 7, 5, 6, 9, 6, 7, 5, 8.$$

a. Set the marks out in the form of a frequency table.
b. Calculate their mean and standard deviation from the frequency table.

3. The scores obtained by individuals in a certain experiment have been found to be normally distributed, with mean 60 and standard deviation of 10.
a. The scores obtained by four individuals were 58, 70, 90 and 52. Express these performances in the form of standard scores.
b. What is the probability of obtaining by chance from this population an individual whose score
  (*i*) Is 67 or greater

(*ii*) Lies between 50 and 64

(*iii*) Is 45 or less

(*iv*) Is either 52 or less, or 72 or more?

c. Student are divided into groups of 4 for this experiment and the mean score for each group determined. How would these group means be distributed?

d. Are we justified in considering the performance of the group in *a.* as unusual?

4.a. Explain succinctly why if two events A and B are mutually exclusive they cannot also be independent.

   b. In a certain community the probability that a home is equipped with a television receiver is 0.5. Explain the fallacy in the following reasoning:

   'If we choose two independent homes at random, at least one must
   have a television receiver, because

   Probability (at least one home has TV)

   = Probability (first home has TV)

      + Probability (second home has TV)

   = 0.5 + 0.5

   = 1.0'

c. Determine the probability in *b.* that exactly one of three houses chosen at random should have a television receiver.

d. For the purpose of a social survey, the community in *b.* has been divided into a large number of groups each containing 100 houses. The number of homes with TV sets is determined for each of these groups. What would be the mean and standard deviation of the data so obtained?

A4

1. Eight spiders of a certain species had their body lengths measured to the nearest millimetre, the results being

15   20   14   14   15   13   17   20

a. Determine

  (*i*) The mean

 (*ii*) The median

(*iii*) The mean deviation

(*iv*) The standard deviation.

b. Express the value 20 as a standard score.

2. A set of data has been condensed into a frequency table as follows:

Class interval	Frequency
100 — under 120	8
120 — under 140	14
140 — under 160	8

a. Sketch the less than ogive for this data.
b. Calculate the standard deviation of the data.
c. Estimate the median.

3. A population of horses has weights approximately normally distributed with a mean of 400 kg and a standard deviation 50 kg.

a. Determine the probability that a randomly-chosen horse should have weight
 (*i*) Between 350 and 525 kg
(*ii*) Greater than 515 kg
b. If a random sample of 16 horses were taken, what could you say about the probability distribution of their average weight?
c. What is the probability that the average weight obtained from such a sample would be less than $362\frac{1}{2}$ kg ?
4. For the following data:

21.5	15.8	19.4	17.2	21.9	27.0	18.5	22.7	23.4	20.3
23.2	22.8	21.2	24.1	16.7	24.2	24.8	20.7	21.3	30.5
20.2	23.9	16.1	17.9	19.9	29.8	21.8	22.4	28.2	24.0

a. Draw up a frequency table with a class size of 2.0, the first class to be 15.0 and up to 16.99.
b. Calculate the sample mean, from the frequency table.
c. Calculate the sample standard deviation.
d. Find the range and the lower quartile.

A5

1. The marks of 10 individuals in a test were as follows:

$$0, 2, 2, 4, 4, 4, 5, 5, 6, 8.$$

a. From these individual marks calculate
 (*i*) The mean mark
 (*ii*) The median
(*iii*) The mean deviation
(*iv*) The standard deviation.

b.(*i*) Arrange the marks in the form of a frequency table.

(*ii*) From the frequency table calculate the mean and the standard deviation.

c. You are told that this test has been given many times in the past and the marks have been normally distributed with mean 5 and standard deviation 1.5. Is there any reason to suspect that this group of 10 individuals is an unusual one? (Justify your answer.)

d. Using this past experience as a basis, express the marks 8 and 4 as standard scores.

2. The number of accidental male deaths in a certain area in a given year were:

Age	Number of deaths
Under 5	50
Over 5 — under 10	70
Over 10 — under 15	160
Over 15 — under 20	670
Over 20 — under 25	450
Over 25 — under 30	100
Over 30	0

a. Draw a less than ogive for this data (do not spend time drawing a smooth curve, just join the points).

b. Determine the median age at death

(*i*) from the ogive

(*ii*) from the above grouped frequency table.

c. Estimate from the ogive the number of male deaths for the age group over 17 — under 22.

3. If marks in an examination are normally distributed with mean 60 and standard deviation 10, what percentage of candidates would have marks between 50 and 75?

What is the chance that the average mark of 16 candidates selected at random is less than 63?

4. A population consists of the four values 3, 6, 9 and 10.

a. Calculate the population mean.

b. Calculate the population standard deviation.

c. List all the possible samples of size 2 which can be selected from the population.

d. Calculate the means of each of these samples.

e. Calculate the standard deviation of the means of the samples.

A6

1. A nurseryman has bred a new variety of hibiscus and (being statistically minded) has measured the diameters of a sample of five of the flowers as follows, all measurements being in millimetres.

$$127 \quad 135 \quad 129 \quad 138 \quad 131$$

a. Calculate from this data
 (*i*) The mean
 (*ii*) The median
(*iii*) The mean deviation
(*iv*) The standard deviation.
b. Express the value 138 as a standard score.

2. A manufacturer of ballpoint pens has conducted an experiment to test the average life of these pens under conditions of constant use. He sampled 400 pens by placing them in a special device which simulated constant writing, and observed how long each pen survived before ceasing to work effectively. He recorded his results in the form of a frequency table as follows:

Time to failure	Number of pens
under $4\frac{1}{2}$ hrs	0
at least $4\frac{1}{2}$ hrs, but less than $5\frac{1}{2}$ hrs	250
at least $5\frac{1}{2}$ hrs, but less than $6\frac{1}{2}$ hrs	100
at least $6\frac{1}{2}$ hrs, but less than $7\frac{1}{2}$ hrs	50
more than $7\frac{1}{2}$ hrs	0

a. On graph paper, sketch the less than ogive for this data.
b. From the ogive, estimate the proportion of pens which failed in less than $6\frac{1}{4}$ hours.
c. Calculate the standard deviation of this data.

3. The number of articles transmitted through a certain internal mail system in a day under ordinary conditions has been found to be approximately normally distributed with a mean of 2500 and a standard deviation of 400.
a. Determine the probability that, in a given day, the number of articles transmitted through this system should be

 (*i*) Between 2400 and 2900
 (*ii*) More than 3150
(*iii*) Less than 2450.

b. If the system were observed over a period of 16 days, state the characteristics of the probability distribution of the *average* number of articles transmitted per day over that period. Hence evaluate the probability that the *average* number transmitted per day over the 16 day period should be more than 2575.

c. Over a four day period towards Christmas, it was found that the average number of articles transmitted per day was 2850. Do you feel that this evidence would enable you to conclude firmly that the system is busier towards Christmas? Justify your answer statistically.

4. In a certain university, it is reliably estimated that one-half of the female students could reasonably be described as *attractive*. In one class containing 256 female students, the lecturer estimates that only 108 are *attractive*. Could he reasonably infer that there is a genuine tendency for unattractive females to enrol in his class? Justify your answer statistically.

B1

1. In a certain hunt involving black and white bunnies the numbers caught were as follows:

	White	Black	Total
Number caught	9	36	45
Not caught	51	84	135
Total	60	120	180

Are white bunnies definitely harder to catch?

2. A supervisor was asked to rank 10 employees in increasing order of daily output. His estimates were then compared with the results from actual performance. The correct order was:

2, 6, 4, 5, 3, 7, 9, 1, 8, 10

The supervisor placed them in order:

3, 5, 4, 6, 2, 8, 10, 1, 7, 9.

Is it likely that such a measure of agreement could have been obtained by chance?

3. We are told that a spelling test involving a large number of words was designed so that the students' marks would have a mean of 5 and a standard deviation of 1.2.

If the mean mark of 64 students in this test was 5.4, could their teacher claim that they were definitely superior in spelling? Would it alter your method of solving the problem if you were told that students' marks are normally distributed? If not, why not?

4. Students of six tutorial classes in english obtained the following marks in a term essay:

Class	Marks
A	23 27 36 39 33 22 31 37 23 40 32 21 34 29 27 30
B	21 28 30 29 34 33 20 37 39 18 31 40 34 22 25 30 28 20
C	17 39 30 23 31 26 18 27 20 38 21 29 38 24 33 19
D	17 31 24 37 19 32 25 20 35 21 36 22 38 26 32 23
E	23 31 15 28 26 18 35 19 36 22 27 19 28 20 29 24 21 25
F	14 28 20 16 33 24 19 24 22 29 22 25 21 30 23 20

The median mark for the six classes combined is $26\frac{1}{2}$. Is it likely that these six classes are random samples from the same population with a median of $26\frac{1}{2}$?

5. A survey of the linguistic habits of school pupils has produced the following numbers using a short vowel and a long vowel in a particular word:

Numbers using a long vowel

	Private school		State school	
	Males	Females	Males	Females
City schools	156	134	215	228
Country schools	102	88	224	233

Numbers using a short vowel

	Private school		State school	
	Males	Females	Males	Females
City schools	207	213	192	231
Country schools	163	237	226	251

Is there any real difference in the proportion using a long vowel in private and in state schools?

6. Twenty students who sat for both mathematics and english obtained the following marks:

Marks in english	*Marks in mathematics*
6	10
8	5
9	6
4	7
8	5
10	7
8	6
5	8
6	8
8	6
5	8
5	8
7	6
5	6
7	6
6	9
8	6
10	7
9	6
6	10

The mean and the standard deviation of the 20 marks for each subject have been calculated and are:

	Mean	SD
English	7	1.73
Mathematics	7	1.45

Calculate the product moment correlation coefficient.

What conclusion do you draw from it?

(You may assume that mathematics and english marks are nearly normally distributed.)

B2

1. The number of left-handed and right-handed teenagers in a certain community who were graded as A or B grade tennis players and the numbers in the same community who were below B grade were as follows:

	A or B grade	Below B grade	Total
Left-handed	18	182	200
Right-handed	82	1,718	1,800
Total	100	1,900	2,000

Can we conclude that left-handers are better tennis players? (Justify your answer.)

2. A certain university course has both theoretical and practical aspects which can be tested separately. Ten students were ranked in the two sections as follows:

Rank in theory test	Rank in practical test
3	1
6	5
4	2
7	10
10	7
1	3
5	8
2	4
9	9
8	6

Is there any significant difference in the performance of the students in the two sections of the course?

3. Of 200 alligators in zoos throughout the world, 25 died in a certain year. Of 150 crocodiles in captivity, 25 died. Is this substantial evidence for a difference in fatality rates between alligators and crocodiles under conditions of captivity?

4. A university divides its students in demography into three classes namely, day, evening and external. The students' marks in the final examinations were:

Class	Marks									
Day	73	87	86	89	83	72	81	87	73	90
Evening	71	78	80	79	84	83	70	77	89	88
External	67	89	80	73	81	76	68	77	70	68

a. What is the median mark for the university?
b. Is it likely that the three classes are random samples from the same population?

5. For the marks for the university in question *4*
a. Draw a less than ogive
b. Determine from the ogive, the median, the upper and lower quartiles and the ninety percentile.

6. A television producer claimed that his programme attracted one third of the viewing public. Out of a random sample of television viewers 30 out of 120 had their sets tuned to his programme. Is this significantly below his claim?

B3

1. It is found that 500 out of the 1,000 university students in a survey smoke cigarettes. There were 500 female students in the survey out of whom 300 smoked. Does this indicate a significant difference in the incidence of smoking between the sexes?

2. One brand of fertilizer applied to a sample tract of 37 hectares resulted in a yield of 33 kg per hectare with a standard deviation of 5 kg. Another brand resulted in a yield of 32.8 kg per hectare on a sample tract of 37 hectares, with a standard deviation of 5 kg.
Is there any significant difference in the yields of the two brands?

3. The results of a survey of reactions to a television programme were as follows:

| | Sex | | |
Reaction	Male	Female	Total
Favourable	28	52	80
Neutral	20	20	40
Unfavourable	52	28	80
Total	100	100	200

Is there a significantly different reaction by the two sexes?

4. Twelve investments are ranked by two analysts in terms of degree of risk as follows:

Investment	Rank by X	Rank by Y
A	7	6
B	8	4
C	2	1
D	1	3
E	9	11
F	3	2
G	12	12
H	11	10
I	4	5
J	10	9
K	6	7
L	5	8

Are the analysts in substantial agreement?

5. Two variables take the following values:

X	Y
1	9
3	7
5	5
7	3

Is there a significant correlation between them? Explain clearly any assumptions on which your decision is based.

6. The manufacturer of a certain product claims that it will last 2500 hours. A sample of 17 had an average life of 2325 hours and a standard deviation of 600 hours. Does this sample indicate that his product does not live up to his claims?
(You may assume that the life of the product is normally distributed.)

B4

1.a. The pass rate in a certain public examination is 80%. Given three students chosen at random who undertake this examination, what is the probability that exactly one of them will fail?

   b. In equivalent examinations in the past, a pattern has emerged whereby the grade distribution was as follows:

1st class honours	5%
2nd class honours	10%
A grade pass at ordinary level	25%
B grade pass at ordinary level	40%
Fail	20%

In 1968, the numbers who gained the various grades were as follows:

1st class honours	209
2nd class honours	412
A grade pass at ordinary level	1020
B grade pass at ordinary level	1564
Fail	795
	———
	4000

Is there any evidence of substantial departure from the previous grade distribution pattern?

2. A particular group of 10 students sat for tests in two subjects, english and mathematics. Their marks were as follows:

Student	English (x)	Mathematics (y)
Smith	9	8
Green	8	8
Jones	8	7
Black	7	9
Robinson	7	1
Brown	6	10
Watson	5	7
Mason	4	4
Miller	4	3
Baker	2	3

a. Considering these tests as ranking procedures, is there significant agreement between the tests?

b. By plotting the above test results on graph paper estimate the product moment correlation between marks in english and mathematics.

c. Given that $\Sigma x = 60$, $\Sigma x^2 = 404$, $\Sigma y = 60$, $\Sigma y^2 = 442$ and $\Sigma xy = 391$, write down (but do not simplify) an expression involving numbers only, for the product moment correlation coefficient.

3.a. What conclusion would you draw from a chi-squared value of 14.21 obtained by comparing the eight actual frequencies in a 2×4 contingency table with frequencies calculated from:

$$\text{Expected frequency of cell in row A and column B} = \frac{\text{Row A total} \times \text{column B total}}{\text{Grand total}}$$

b. A product moment correlation coefficient of -0.66 was obtained from a sample of 28 pairs of x and y values. Is this result significantly different from no real correlation?

c. The mean of a sample of 100 (from a normal distribution with a mean of 20 and a standard deviation of 4) is 21. What is the probability of obtaining a sample whose mean differs by as much as this from the population mean?

d. What would your answer be to *c.* if the population standard deviation was unknown and the sample standard deviation $s = 5.1$? Explain.

4. A group of ten students sat for two tests in a certain subject. Their results were:

Test \ Students	1	2	3	4	5	6	7	8	9	10
A	51	40	88	60	57	72	65	55	58	45
B	51	44	87	59	59	74	68	57	62	50

Use the appropriate *t*-test to find if there is any significant difference between the students' results on the two tests.

5.a. What basic assumption was made in the test in question *4.* above?

b. Use another test which does not involve an assumption of this nature, to test whether there is significant difference between the students' results in the two tests given in question *4.* above.

B5

1. The supervisor at a manual telephone exchange has noted over many years that the time it takes to connect a call has a mean of 35 seconds, with a standard deviation of 9 seconds. A new well-trained operator has just come on duty, and the supervisor notes that she has connected 36 calls in an average time of only 31 seconds per call. Do you feel confident that this new operator is really more proficient than her predecessors? (Justify your answer statistically.)

2. The scores of high school students in a well-established test of motor skills are known to be normally distributed with a mean of 60 and a standard deviation of 10.

a. What proportion of the population have scores between 54 and 66?

b. How many students in a group of 200 would be expected to have scores between 60 and 65?

c. What is the chance that 2 students selected at random will both have scores above 65?

d. In a small class of 25 the average score was 55. Is this unusual?

3.a. A paint manufacturer markets 10 different colours of interior wall paint. The weekly sales volumes (from which a consumer preference ranking can be inferred) are given below. The magazine *Interior Decor* has surveyed current trends and has ranked these same colours in a different order. Is there a significant relationship between the two rankings?

Weekly sales volume		*Interior Decor* ranking
Colour	Gallons	
aqua	100	off-white (most popular)
bone	300	white
cream	280	cream
daffodil	210	daffodil
grey	85	bone
off-white	800	pink
pale blue	220	pale blue
pale green	200	grey
pink	250	aqua
white	1020	pale green (least popular)

b. A product moment correlation coefficient of —0.73 was obtained from a sample of 7 pairs of observations of average speed (on country runs) and petrol consumption. Does this indicate a correlation significantly different from zero?

4. A group of full-time university students were interviewed in regard to their living arrangements and their prospects of graduating in minimum time. The following frequencies were obtained:

	Student's estimate of chance of graduating in minimum time		
	Poor	Fair	Good
Living at home	53	102	45
Living away from home	7	18	15

Does this indicate a statistically significant relationship between the two classifications? Comment on possible interpretations of the result.

5. As a statistician, you are given the following data, without explanation, and are asked to determine whether or not a significant difference exists between the two situations under study.

Performance in manual dexterity test

Situation \ Subject No.	1	2	3	4	5	6	7	8	9	10
1. Natural light and ventilation (x_1)	26	42	31	53	24	33	37	40	22	38
2. Artificial light and air-conditioning (x_2)	24	35	32	42	26	34	34	34	21	34

Discuss the assumptions you could make and the tests that could be used if those assumptions were valid. Compare the results of at least 2 possible tests. If required in the course of your answer, you may use the values

$$\bar{x}_1 = 34.6 \qquad \bar{x}_2 = 31.6$$
$$s_1 = 8.95 \qquad s_2 = 5.87$$

B6

1. Five hundred university students were asked whether they believed that the Sydney water supply should be fluoridated without a referendum being held. Each student interviewed was placed into one of the categories 'arts', 'economics' or 'science' depending on his major field of study. The responses in the survey were summarised as follows:

	Arts	Economics	Science
Fluoridate unconditionally	50	42	58
Hold referendum	150	58	142

Can you detect any significant difference in outlook between the students in the various fields?

2. A railway executive has been investigating whether there is any tendency for passengers to travel in the upper deck of double deck carriages in preference to the lower deck. He has collected data by selecting ten trains at random during periods when the cars are neither crowded nor near-empty, with results as follows:

Observation number	1	2	3	4	5	6	7	8	9	10
Number travelling in upper deck	37	42	33	32	35	29	33	37	38	34
Number travelling in lower deck	34	43	25	38	34	24	27	30	36	29

Perform two separate tests, one making an assumption that the data satisfies normality conditions and the other without such assumption, to determine whether there is any significant preference exhibited for travel on the upper deck.

Comment also on the plausibility of the normality assumption.

3. An ecologist conjectures that there is a connection between the dawn temperature and the time taken for an animal to bestir after dawn. He has observations on a set of farm animals over a seven-day period, as follows:

Day	Sun.	Mon.	Tues.	Wed.	Thur.	Fri.	Sat.
Dawn temperature (°C) (x)	4	5	6	8	8	7	5
Average time for animals to become active (minutes) (y)	23	20	22	16	15	24	27

On the basis of this evidence, do you feel there is substantial justification for the ecologist's conjecture? Justify your answer statistically, once again using one method without a normality assumption and one method with such assumption.

(You are advised to give consideration to choice of calculating methods with minimum arithmetic complexity. If required for the methods you choose, you may use the facts that

$$\Sigma x = 43, \quad \Sigma y = 147, \quad \Sigma x^2 = 279, \quad \Sigma y^2 = 3199.)$$

4. A music critic has frequently complained that some record companies have a tendency to provide less playing time per record than others. He has been challenged by the companies to establish his complaint by statistical means. He has therefore collected the following data by observing the total playing time of ten records of popular music from each of four firms, as follows:

Company					Playing times					
A	45	32	38	39	28	38	37	47	32	30
B	28	32	32	32	27	25	26	29	34	38
C	31	28	38	39	40	41	32	28	40	36
D	36	42	44	40	38	39	50	36	32	30

Is any significant difference amongst the companies established by these figures?

5. A psychologist has defined a quantity known as a 'loquacity index' which indicates the extent to which an individual uses more words than necessary to convey an idea. He has discovered from large-sample experimentation that this index is distributed around an average of 50 with a standard deviation of 30, and that observed values of the index range from zero to 200.

a. Use normal distribution methods to estimate the probability that the average loquacity index of 81 randomly chosen people should lie between 46 and 53.

b. How does the statement of the situation make it clear that loquacity indices could *not* be regarded as normally distributed?

c. Why then was it legitimate to make use of the normality operations in *a.* even though the parent distribution was specifically known to be non-normal?

d. The loquacity index was measured for each of 100 trainee public service clerks, and the average index of the group was found to be 55.1. Is this convincing evidence that the trainee clerks have a loquacity pattern different from the population as a whole?

4: Statistical Tables

Areas under the
Standard Normal Curve

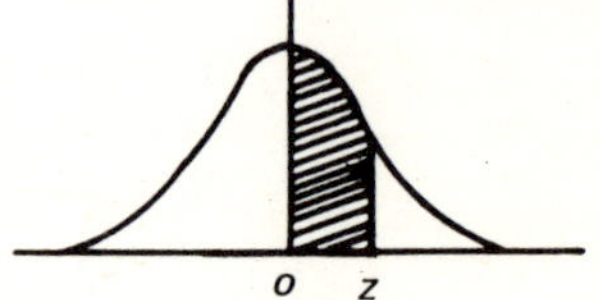

z	·00	·01	·02	·03	·04	·05	·06	·07	·08	·09
0·0	·0000	·0040	·0080	·0120	·0160	·0199	·0239	·0279	·0319	·0359
0·1	·0398	·0438	·0478	·0517	·0557	·0596	·0636	·0675	·0714	·0754
0·2	·0793	·0832	·0871	·0910	·0948	·0987	·1026	·1064	·1103	·1141
0·3	·1179	·1217	·1255	·1293	·1331	·1368	·1406	·1443	·1480	·1517
0·4	·1554	·1591	·1628	·1664	·1700	·1736	·1772	·1808	·1844	·1879
0·5	·1915	·1950	·1985	·2019	·2054	·2088	·2123	·2157	·2190	·2224
0·6	·2258	·2291	·2324	·2357	·2389	·2422	·2454	·2486	·2518	·2549
0·7	·2580	·2612	·2642	·2673	·2704	·2734	·2764	·2794	·2823	·2852
0·8	·2881	·2910	·2939	·2967	·2996	·3023	·3051	·3078	·3106	·3133
0·9	·3159	·3186	·3212	·3238	·3264	·3289	·3315	·3340	·3365	·3389
1·0	·3413	·3438	·3461	·3485	·3508	·3531	·3554	·3577	·3599	·3621
1·1	·3643	·3665	·3686	·3708	·3729	·3749	·3770	·3790	·3810	·3830
1·2	·3849	·3869	·3888	·3907	·3925	·3944	·3962	·3980	·3997	·4015
1·3	·4032	·4049	·4066	·4082	·4099	·4115	·4131	·4147	·4162	·4177
1·4	·4192	·4207	·4222	·4236	·4251	·4265	·4279	·4292	·4306	·4319
1·5	·4332	·4345	·4357	·4370	·4382	·4394	·4406	·4418	·4429	·4441
1·6	·4452	·4463	·4474	·4484	·4495	·4505	·4515	·4525	·4535	·4545
1·7	·4554	·4564	·4573	·4582	·4591	·4599	·4608	·4616	·4625	·4633
1·8	·4641	·4649	·4656	·4664	·4671	·4678	·4686	·4693	·4699	·4706
1·9	·4713	·4719	·4726	·4732	·4738	·4744	·4750	·4756	·4761	·4767
2·0	·4772	·4778	·4783	·4788	·4793	·4798	·4803	·4808	·4812	·4817
2·1	·4821	·4826	·4830	·4834	·4838	·4842	·4846	·4850	·4854	·4857
2·2	·4861	·4864	·4868	·4871	·4875	·4878	·4881	·4884	·4887	·4890
2·3	·4893	·4896	·4898	·4901	·4904	·4906	·4909	·4911	·4913	·4916
2·4	·4918	·4920	·4922	·4925	·4927	·4929	·4931	·4932	·4934	·4936
2·5	·4938	·4940	·4941	·4943	·4945	·4946	·4948	·4949	·4951	·4952
2·6	·4953	·4955	·4956	·4957	·4959	·4960	·4961	·4962	·4963	·4964
2·7	·4965	·4966	·4967	·4968	·4969	·4970	·4971	·4972	·4973	·4974
2·8	·4974	·4975	·4976	·4977	·4977	·4978	·4979	·4979	·4980	·4981
2·9	·4981	·4982	·4982	·4983	·4984	·4984	·4985	·4985	·4986	·4986
3·0	·4987	·4987	·4987	·4988	·4988	·4989	·4989	·4989	·4990	·4990
3·1	·4990	·4991	·4991	·4991	·4992	·4992	·4992	·4992	·4993	·4993
3·2	·4993	·4993	·4994	·4994	·4994	·4994	·4994	·4995	·4995	·4995
3·3	·4995	·4995	·4996	·4996	·4996	·4996	·4996	·4996	·4996	·4997
3·4	·4997	·4997	·4997	·4997	·4997	·4997	·4997	·4997	·4998	·4998

Student's t Distribution[1]

Values of t which are exceeded numerically
with given probability for various degrees
of freedom.

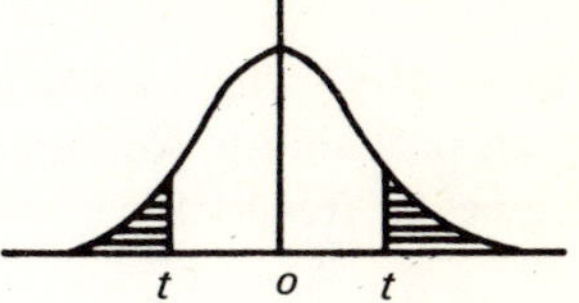

Degrees of freedom	Probability				
	0·10	0·05	0·025	0·01	0·005
1	6·31	12·71	25·45	63·66	127·32
2	2·92	4·30	6·21	9·92	14·09
3	2·35	3·18	4·18	5·84	7·45
4	2·13	2·78	3·50	4·60	5·60
5	2·02	2·57	3·16	4·03	4·77
6	1·94	2·45	2·97	3·71	4·32
7	1·89	2·36	2·84	3·50	4·03
8	1·86	2·31	2·75	3·36	3·83
9	1·83	2·26	2·69	3·25	3·69
10	1·81	2·23	2·63	3·17	3·58
11	1·80	2·20	2·59	3·11	3·50
12	1·78	2·18	2·56	3·05	3·43
13	1·77	2·16	2·53	3·01	3·37
14	1·76	2·14	2·51	2·98	3·33
15	1·75	2·13	2·49	2·95	3·29
16	1·75	2·12	2·47	2·92	3·25
17	1·74	2·11	2·46	2·90	3·22
18	1·73	2·10	2·45	2·88	3·20
19	1·73	2·09	2·43	2·86	3·17
20	1·72	2·09	2·42	2·85	3·15
30	1·70	2·04	2·36	2·75	3·03
40	1·68	2·02	2·33	2·70	2·97
60	1·67	2·00	2·30	2·66	2·91
120	1·66	1·98	2·27	2·62	2·86
∞	1·64	1·96	2·24	2·58	2·81

[1] Modification of a more extensive table by C. M. Thompson, 'Table of percentage points of the incomplete beta function', *Biometrika*, vol. 32, 1941, p.151. By kind permission of the author and publishers.

The χ^2 Distribution[2]

Values of χ^2 which are exceeded with given probability, for various degrees of freedom.

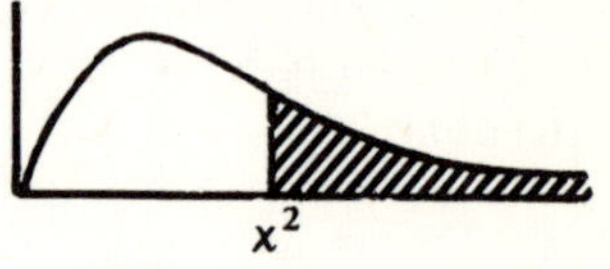

Degrees of freedom	Probability			
	0·050	0·025	0·010	0·005
1	3·84	5·02	6·63	7·88
2	5·99	7·38	9·21	10·60
3	7·81	9·35	11·34	12·84
4	9·49	11·14	13·28	14·86
5	11·07	12·83	15·09	16·75
6	12·59	14·45	16·81	18·55
7	14·07	16·01	18·48	20·28
8	15·51	17·53	20·09	21·96
9	16·92	19·02	21·67	23·59
10	18·31	20·48	23·21	25·19
11	19·68	21·92	24·73	26·76
12	21·03	23·34	26·22	28·30
13	22·36	24·74	27·69	29·82
14	23·68	26·12	29·14	31·32
15	25·00	27·49	30·58	32·80
16	26·30	28·85	32·00	34·27
17	27·59	30·19	33·41	35·72
18	28·87	31·53	34·81	37·16
19	30·14	32·85	36·19	38·58
20	31·41	34·17	37·57	40·00
25	37·65	40·65	44·31	46·93
30	43·77	46·98	50·89	53·67
40	55·76	59·34	63·69	66·77
50	67·50	71·42	76·15	79·49
60	79·08	83·30	88·38	91·95
70	90·53	95·02	100·43	104·22
80	101·88	106·63	112·33	116·32
90	113·15	118·14	124·12	128·30
100	124·34	129·56	135·81	140·17

[2] Modification of a more extensive table by C. M. Thompson, 'Table of percentage points of the χ^2 distribution', *Biometrika*, vol. 32, 1941, p.187. By kind permission of the author and publishers.

Values[3] of $w = \frac{1}{2}\log_e \dfrac{1+r}{1-r}$

r	w	r	w	r	w	r	w
·00	·000	·25	·255	·50	·549	·75	·973
·01	·010	·26	·266	·51	·563	·76	·996
·02	·020	·27	·277	·52	·576	·77	1·020
·03	·030	·28	·288	·53	·590	·78	1·045
·04	·040	·29	·299	·54	·604	·79	1·071
·05	·050	·30	·310	·55	·618	·80	1·099
·06	·060	·31	·321	·56	·633	·81	1·127
·07	·070	·32	·332	·57	·648	·82	1·157
·08	·080	·33	·343	·58	·662	·83	1·188
·09	·090	·34	·354	·59	·678	·84	1·221
·10	·100	·35	·365	·60	·693	·85	1·256
·11	·110	·36	·377	·61	·709	·86	1·293
·12	·121	·37	·388	·62	·725	·87	1·333
·13	·131	·38	·400	·63	·741	·88	1·376
·14	·141	·39	·412	·64	·758	·89	1·422
·15	·151	·40	·424	·65	·775	·90	1·472
·16	·161	·41	·436	·66	·793	·91	1·528
·17	·172	·42	·448	·67	·811	·92	1·589
·18	·182	·43	·460	·68	·829	·93	1·658
·19	·192	·44	·472	·69	·848	·94	1·738
·20	·203	·45	·485	·70	·867	·95	1·832
·21	·213	·46	·497	·71	·887	·96	1·946
·22	·224	·47	·510	·72	·908	·97	2·092
·23	·234	·48	·523	·73	·929	·98	2·298
·24	·245	·49	·536	·74	·950	·99	2·647

[3] P. G. Hoel, *Elementary Statistics*, Wiley, New York, 1960. Reprinted with kind permission of the author and publishers.

Rank Correlation Coefficient[4]

n	Significance level (one-sided test)	
	·05	·01
4	1·000	
5	·900	1·000
6	·829	·943
7	·714	·893
8	·643	·833
9	·600	·783
10	·564	·746
12	·504	·701
14	·456	·645
16	·425	·601
18	·399	·564
20	·377	·534
22	·359	·508
24	·343	·485
26	·329	·465
28	·317	·448
30	·306	·432

[4] E. G. Olds, 'Distribution of sums of squares of rank differences for small numbers of individuals', *Annals of Mathematical Statistics*, vol. 9, 1938, p.133. By kind permission of the author and editor.

Index